*Climatic Fluctuations and Water Management*

D1631708

# Climatic Fluctuations and Water Management

*Edited by*

*Mahmoud A. Abu-Zeid*
*Chairman, Water Research Center*
*Cairo, Egypt*

*and*

*Asit K. Biswas*
*President, International Water Resources*
*Association, Oxford, UK*

**B**UTTERWORTH
**H**EINEMANN

Butterworth-Heinemann Ltd
Linacre House, Jordan Hill, Oxford OX2 8DP

 PART OF REED INTERNATIONAL BOOKS

OXFORD   LONDON   BOSTON
MUNICH   NEW DELHI   SINGAPORE   SYDNEY
TOKYO   TORONTO   WELLINGTON

First published 1992

**British Library Cataloguing in Publication Data**
Abu-Zeid, Mahmoud A.
  Climatic fluctuations and water management.
  I. Title   II. Biswas, Asit
  551.57

ISBN 0 7506 1320 3

**Library of Congress Cataloguing in Publication Data**
Climatic fluctuations and water management/edited by Mahmoud A.
  Abu-Zeid and Asit K. Biswas.
  p. cm.
  Includes bibliographical references and index.
  ISBN 0 7506 1320 3
  1. Hydrometeorology.   2. Climatic change.   I. Abu-Zeid, Mahmoud
  A.   II. Biswas, Asit K.
  GB2805.C58   1992                                          91–36159
  551.57–dc20                                                CIP

Printed and bound in Great Britain by Billings & Sons Ltd, Worcester

# CONTENTS

# PREFACE

The issue of climatic change has become a major environmental concern on a global basis in less than a decade. If one reviews the landmark United Nations Conference on the Human Environment held in Stockholm in 1972, the main environmental concerns of the world, as expressed in its Action Plan, were in terms of air and water pollution, loss of productive soil, resource depletion, deforestation, and other similar issues. The potential problem of global warming due to increasing concentrations of greenhouse gases was not regarded as sufficiently important to merit any deliberation. Thus, within a short time scale of less than two decades, the issue of global warming has advanced from being virtually unkown to the centre of world attention.

If climatic change contributes to the occurrence of many catastrophic events of magnitudes and scales that scientists are predicting, then unquestionably it would be one of the most serious problems man has ever had to face and, hopefully, resolve within the short time span of only a few decades. It could have the potential to change significantly patterns of human settlements, agricultural production, energy production and use, and many other interrelated factors. In the area of water resources, it could alter precipitation regimes, river flows, rates of groundwater recharge as well as water quality. Sea levels could rise and affect low-lying coastal areas.

It is possible that all the above-mentioned events could occur, and because of the scale and magnitude of their potential impacts, we must take them seriously. Unfortunately, however, at our present state of knowledge all these can, at best, be considered scenarios that remain hidden in a series of interconnected uncertainties. Not only is it difficult to forecast the rates of increase in the various greenhouse gases over the next few decades it is also even more complex to translate the results of these increases into real impacts with any degree of accuracy. The global circulation models available at present have many shortcomings, ranging from the merely conceptual, such as representations of clouds and atmosphere–ocean interactions, to shortages of long-term reliable data on the oceans. Accordingly, forecasts must be considered to be very tentative in nature. Nor do the models predict consistent patterns of regional climatic changes.

Even if the predictions made by the global circulation models were considered to be reliable, they would still, at present, be of minor interest in terms of planning, design and operation of specific individual water projects. This is because the current generation of models simply cannot provide usable climatic information on specific river basins and irrigated agricultural project areas on a scale that can be effectively used by water professionals. Equally, simple and general information on average annual rainfall and runoff in an area are inadequate for management of water projects. Hydrological characteristics of water projects must consider short-time-scale events such as floods or 10-day low flows for design and operation. Such micro-scale and detailed forecasts are not available from any of the current global circulation models.

In contrast to the issue of climatic change, which is beset with uncertainties at present, climatic fluctuation is a well-known phenomenon. We now have several decades of hydroclimatic records from different parts of the world which clearly indicate that the extent of problems posed by climatic fluctuations have been somewhat underestimated in the past. The magnitude and extent of the problem could vary from country to country, or even from one part of a large river basin to another. During recent years, an important research focus has been an attempt to predict climatic changes that could occur in the future (say, between the years 2000 and 2050). This concentration of research emphasis primarily on potential future climatic changes has, unfortunately, appeared to have diverted the attention of hydrologists, climatologists and other water professionals from the urgent necessity of dealing effectively with climatic fluctuations which have already been observed in many river basins over the past decades, and thus may occur in the future. These fluctuations are not merely a hypothesis: they are a reality.

In order to emphasize the importance of climatic fluctuations in water management, the International Water Resources Association, Water Research Center of the Ministry of Public Works and Water Resources, Government of Egypt, and the United Nations Environment Programme co-sponsored an International Seminar on Climatic Fluctuations and Water Management in Cairo, Egypt, on 11–14 December 1989. The Seminar was attended by 150 participants from 38 countries and many international organizations. Seventy papers were received for presentation and discussion.

This volume includes a selected group of papers that were presented during the Cairo Seminar. It is hoped that the Seminar and its follow-up activities will contribute to the realization of the present unsatisfactory state of knowledge of how to deal effectively with climatic fluctuations within the context of water development and management. This, in turn, could then lead to accelerated research in this area, which, hopefully, would provide water professionals with more efficient methodologies for dealing with such complex problems. This was our long-term goal when we decided to convene the Seminar. We hope that the present volume will contribute to the discussions and further accelerate research in this area, which will go a long way towards achieving this goal.

*Mahmoud A. Abu-Zeid*
*Asit K. Biswas*

# CONTRIBUTORS

**S. Abdel-Dayem**
Secretary-General, Water Research Center, 22 El-Galaa Street, Bulaq, Cairo, Egypt

**W. A. Abderrahman**
Division of Water Resources and Environment, The Research Institute, King Fahd University of Petroleum and Minerals, Dhahran 31261, Saudi Arabia

**A. B. Abulhoda**
Ministry Consultant, Ministry of Public Works and Water Resources, Cairo, Egypt

**M. A. Abu-Zeid**
Chairman, Water Research Center, 22 El-Galaa Street, Bulaq, Cairo, Egypt

**F. F. Al-Muttair**
Civil Engineering Department, King Saud University, PO Box 800, Riyadh 11421, Saudi Arabia

**A. S. Al-Turbak**
Civil Engineering Department, King Saud University, PO Box 800, Riyadh 11421, Saudi Arabia

**N. Arnell**
Institute of Hydrology, Wallingford, Oxford OX10 8BB, UK

**B. B. Attia**
General Director, Water Research Center, Ministry of Public Works and Water Resources, Cairo, Egypt

**J. Balek**
UNEP, Nairobi, Kenya

**M. A. Bin Afeef**
Department of Meteorology, Faculty of Meteorology, Environment and Arid Land Agriculture, King Abdulaziz University, Jeddah

**A. K. Biswas**
President, International Water Resources Association, 76 Woodstock Close, Oxford OX2 8DD, UK

**P. Bruneau**
Hydro-Québec, Service Hydraulique, Division Hydrologie, 855 Ste-Catherine est, Montreal, QC H2L 4P5, Canada

**J. M. Byrne**
Coordinator, Water Resources Institute, University of Lethbridge, Lethbridge, Alberta, Canada

**S. H. Fahmy**
Chairman, Egyptian Public Authority for Drought Projects, Cairo, Egypt

**M. Fricke**
Hydrogeo Consulting Ltd, Bad Driburg, Germany

**O. E. Frihy**
Water Research Center, Coastal Research Institute, 15 El-Pharaana Street, El-Shallalat, Alexandria, Egypt

**P. H. Gleick**
Pacific Institute for Studies in Development, Environment and Security, Berkeley, USA

**M. Hulme**
Climatic Research Unit, University of East Anglia, Norwich, UK

**C. H. Hulsbergen**
Harbours, Coasts and Offshore Technology Division, Delft Hydraulics, PO Box 152, 8300 AD Emmeloord, The Netherlands

**Ph. Kandilis**
University of Athens, Greece

**S. N. Kathuria**
Central Water Commission, New Delhi, India

**A. A. Khafagy**
Water Research Center, Coastal Research Institute, 15 El-Pharaana Street, El-Shallalat, Alexandria, Egypt

**A. Khan**
Division of Water Resources and Environment, The Research Institute, King Fahd University of Petroleum and Minerals, Dhahran 31261, Saudi Arabia

**W. Klohn**
Food and Agriculture Organization of the United Nations, Via delle Terme di Caracalla, 00100 Rome, Italy

**S. D. Kulkarni**
Central Water Commission, New Delhi, India

**A. M. Mahar**
Water Research Center, Coastal Research Institute, 15 El-Pharaana Street, El-Shallalat, Alexandria, Egypt

**H. T. Mantis**
Research Centre for Atmospheric Physics and Climatology, Academy of Athens, Athens, Greece

**R. B. McNaughton**
Assistant Professor, Department of Geography, University of Lethbridge, Lethbridge, Alberta, Canada

**D. A. Metaxas**
University of Athens, Greece

**A. M. Negm**
Faculty of Engineering, Zagazig University, Egypt

**T. M. Owais**
Faculty of Engineering, Zagazig University, Egypt

**E. Radi**
Minister, Ministry of Public Works and Water Resources, Cairo, Egypt

**C. C. Repapis**
Research Centre for Atmospheric Physics and Climatology, Academy of Athens, Athens, Greece

**P. J. Reynolds**
28 Elvaston Avenue, Nepean, Ontario K2G 3T4, Canada

**W. E. Riebsame**
Department of Geography and Natural Hazards Research and Applications Information Center, University of Colorado, Boulder, CO 80309, USA

**D. Schaffer**
Research Assistant, Water Resources Institute, University of Lethbridge, Lethbridge, Alberta, Canada

**S. Shalash**
Water Research Center, Cairo, Egypt

**S. H. Sharaf El-Din**
Water Research Center, Coastal Research Institute, 15 El-Pharaana Street, El-Shallalat, Alexandria, Egypt

**A. A. Shata**
Professor Emeritus, Desert Institute, Cairo, Egypt

**K. M. Strzepek**
Center for Advanced Decision Support for Water and Environmental Systems, University of Colorado, Boulder, CO 80309–0428, USA

**Y. Takahasi**
Professor, Shibaura Institute of Technology, Tokyo, Japan

**H. Tamiya**
Department of Geography, Faculty of Letters and Education, Ochanomizu University, 2-1-1 Otsuka, Bunkyou-ku, Tokyo 112, Japan

**M. K. Tolba**

Executive Director, UNEP, Nairobi, Kenya

**J. B. Valdes**
Civil Engineering Department and Climate Systems Research Program, Texas A&M University, College Station, TX 77843–3136, USA

**S. Weyman**
Hydro-Québec, Service Hydraulique, Division Hydrologie, 855 Ste-Catherine est, Montreal, QC H2L 4P5, Canada

# Part 1

# Monitoring, Forecasting and Analysis Procedures

# 1 CLIMATIC FLUCTUATION AND PRECIPITATION CHANGE

Hyoe Tamiya
*Ochanomizu University, Tokyo, Japan*

## 1.1 Introduction

At present it appears inevitable that the anthropogenic increase in atmospheric concentration of carbon dioxide and other greenhouse gases will cause global warming in the next century, or even within the next decade.

There are still many uncertainties about the reality of global warming, but it seems very important to estimate the future physical situation and to take countermeasures. Our first concern is change in precipitation conditions, because man's existence depends on water. The purpose of this chapter is to review problems concerning the change in precipitation caused by global warming from a climatological viewpoint.

If the global temperature rises, then the amount of precipitation must also increase due to acceleration in the water cycle. This, however, will take place only on a global scale and over a corresponding longer period. We have no information on future changes in geographical distribution, seasonal variation and intensity of precipitation. The basic parameters for water use remain completely unknown.

We have many climatological maps or data collections on precipitation for the mean condition over a period of more than 30 years. These are useful for long-range planning but not for water management over a shorter time span. For this we need parameters such as the frequencies of occurrence of amounts of anomalous precipitation.

## 1.2 Anomalous precipitation

To examine these parameters, the Japan Meteorological Agency processed world precipitation data (JMA, 1989) and used monthly information from 131 stations (Figure 1.1). These stations were selected because they possess reliable data dating from 1900 as well as information on geographical distribution. They are divided into seven regions: North America (20 stations), South America (19), Europe (24), Africa (17), Soviet-North Asia

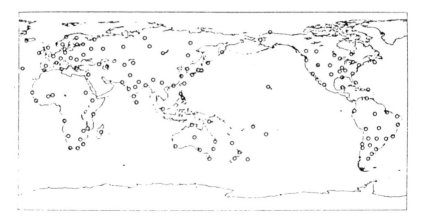

. **Figure 1.1**   *Map of stations used for the study of anomalous precipitation (JMA, 1989)*

(20), South Asia (11) and Oceania (10). The other 10 stations are used only for global study.

Amounts of anomalous high and low precipitation are determined for every decade since 1930 as follows. First, the cubic roots of monthly amounts of precipitation are obtained. It can then be assumed that the frequency of precipitation amounts is normalized. The normals for each month are determined from the data for every station and decade. A normal is defined by the mean value of the previous three decades, i.e. for the 1930s (1930–1939) the normal values are determined from the period 1900–1929. Using the normal and the standard deviation of the same period, all data are checked to determine whether they are within the upper (for high precipitation) and lower (for low precipitation) limits at the significance level of 5%. The number of cases exceeding 5% are totalled as anomalous high or low precipitation. For the 1980s (1980–1987) totalled values are multiplied by 10/8.

The results are given in Figure 1.2 as a mean occurrence number of anomalous amounts of high and low precipitation and the sum of both extremes for a station. Thus 3 (2.5%) is a critical value for the high or low anomalous amounts, and for the sum of both extremes, 6 (5%) is critical.

In some cases, these parameters may also be indicative of long-term climatic fluctuations. As the climate changes or fluctuates, the mean values and the variability of meteorological parameters (in this case, monthly precipitation amounts) may also change. If the mean value of a meteorological parameter moves to high (or low), the probability of occurrence of higher (or lower) values compared to the normal determined from the data of the previous three decades may increase. If the change in climate follows the increase in variance, then probability of anomalous values may increase either way (high or low).

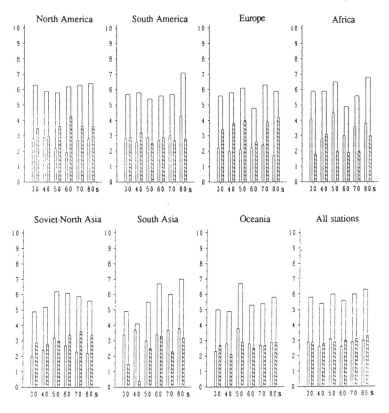

**Figure 1.2**   *Occurrence of anomalous monthly precipitation (JMA, 1989), Narrow hatched columns – low precipitation: narrow plain columns – high precipitation; wide plain columns – sum of low and high*

From the results it is clear that globally (all stations) the amounts of anomalous precipitation have shown no significant changes over the past six decades, but every region has characteristic features. In North America the occurrences of anomalous low precipitation are more than 3 and of high less than 3 (except the 1940s). The sum of both shows no significant change, which means a shift of the mean precipitation amount to the lower value without changes in variability. The shift is most apparent in the 1960s. Nevertheless, the 'Dust Bowl' caused by prolonged drought in the 1930s cannot be indicated by this technique. In the 1980s the anomalous high precipitation increases considerably in South America but there is no clear change in the low precipitation. This suggests an increase in variation over the decade. In Europe the occurrences of anomalous amounts of low precipitation is more than 3, except in the 1960s, and the high is consistently less than 3. This may correspond to a precipitation decrease, for there is no obvious change in the total except in the 1960s. Anomalous high precipitation is found in Africa for four of the six decades. Occurrences of anomalous low precipitation are mostly less than 3. The sum of both extremes changes after the 1960s from less than 6 to more than 6. There are no indications of the recent drought in West Africa, which suggests limitations in this procedure. After the 1960s the low precipitation increased and the high precipitation decreased in Soviet-North Asia. A shift of the mean value to a lower one is to be expected. In South Asia anomalous high precipitation is found for all six decades. Occurrences of anomalous low precipitation are few in the 1930s and 1940s and in recent decades increases in the total are noticeable. Except for the occurrence of anomalous high precipitation and the total of extremes in the 1950s, deviations from the normal are small in Oceania.

The regional aspects described above are interesting when compared with the global one, which shows no significant changes. They are, however, problematic, because assumptions resulting from the normalization of precipitation frequency by using the cubic root are often false. Whether the normal obtained from the mean of three decades is adequate needs further examination. The major problem of lack of data will be discussed in the next section.

## 1.3 Lack of data

The most serious problem for analysis of precipitation is the lack of data, as precipitation shows high spatial variability. In contrast, the temperature distribution shows smaller variability. The results for temperature using a similar procedure are given in Figure 3.1. The increase in the anomalous high temperature in the 1930s and 1980s and in the anomalous low one from the 1950s to the 1970s correspond to a change in the global mean surface temperature or in that for the Northern Hemisphere (Figure 1.4). As yet we do not have such a hemispheric mean for the precipitation amount.

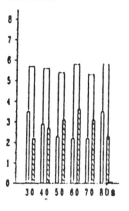

**Figure 1.3** *Occurrence of anomalous monthly temperature for all 131 stations (JMA, 1989). (Key as for Figure 1.2)*

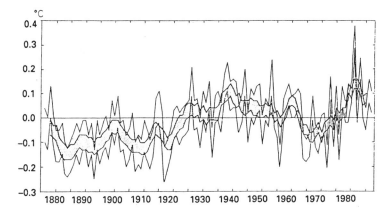

**Figure 1.4** *Secular change of annual surface temperature anomaly in the Northern Hemisphere (1876–1987). The two lines show a 68% confidence interval of estimated error and the shaded areas the same for a 5-year running mean*

This lack of precipitation data must be overcome to deal with problems of regional water management as well as global climatic change. As the regional precipitation condition is a part of the global climate, immediate establishment of a global precipitation measurement network is envisaged.

However, construction of an adequate network is difficult in sparsely populated areas. For oceans, the problem is even greater. Fixed observation platforms, weather ships or anchored buoys are expensive and difficult to maintain. The use of commercial vessels is unsatisfactory, as their routes are limited. Measurements by rain gauges on a rolling platform are another problem. Improvements in these facilities will gradually take place but perhaps not in time, if global warming occurs.

The possibility remains of measurements by satellites. There are some projects being planned or under way: GPCP (Global Precipitation Climatology Programme) and TRMM (Tropical Rainfall Measuring Mission), which are parts of WCRP (World Climate Research Programme) being developed by WMO and ICSU.

GPCP is a project to determine precipitation amounts using cloud height data (infrared radiance) obtained by geostationary and polar orbiting meteorological satellites now in operation (WMO, 1988b). The problems are data collection and organization, and the development of the method to transform the data into precipitation amounts. The overall scheme has been established and is now in the phase of verification using test data, which are provided by JMA. JMA has a comprehensive rain gauge network called AMeDAS (Automatic Meteorological Data Acquisition System). AMeDAS has, on average, one rain gauge every 17 $km^2$.

TRMM plans to place a rain radar and other radiometers on a satellite with an orbital inclination of 30° with the sun asynchronous. This can obtain important tropical precipitation data. The most important aspect is that a new satellite must be launched, and implementation of the plan is expected in the mid-1990s (Theon and Fugono, 1988). At that time, GEWEX (Global Energy and Water Cycle Experiment) will be started, as the largest subprogramme of WCRP (WMO, 1988a).

Perhaps in the next century, our knowledge of climate, including precipitation conditions, will reach the stage of being most useful for the problems concerning water management.

## 1.4  Outlook for the future

Before then, however, it is better to have information on the change in precipitation conditions corresponding to a climatic change by assumed global warming. There are two possibilities: empirical estimation from past data and simulation using a general circulation model (GCM).

Wigley *et al.* (1980) obtained the precipitation difference between the warmest 5 years and the coldest 5 years (Figure 1.5) for the 50 years since 1925 for the Northern Hemisphere using data from 215 stations. The estimation may be applicable to the initial stage of warming, when the boundary conditions are similar to those at present. Kellogg and Schware (1981) used geological or paleoclimatic data and reconstructed a map of global distribution of soil moisture in the Altithermal, 4500–8000 BP (Figure 1.6). In this method, the low reliability and lack of data are disadvantages. The estimation may be applicable to the mature stage of warming, when boundary conditions are considerably changed. Both empirical methods give the same indications for the north-eastern part of the African continent, namely, precipitation increase and greater soil moisture. However, for Japan, the indications are opposite. As the authors point out, results from the empirical method are to be regarded only as scenarios.

Many GCMs are now in operation worldwide and have been used for experimental simulation of increased carbon dioxide concentrations. The general features of the experiments, such as global mean temperature, zonally averaged precipitation distribution, etc., are similar, but regional or local features vary from one model to another. The results of such model simulations are very informative for research purposes. One recent experiment in Japan (Noda and Tokioka, 1989) shows that a decrease in precipitation area and an increase in precipitation intensity occur under doubled carbon dioxide conditions but only in the global or hemispheric domain. In actual use, however, such generalized information cannot be applicable effectively.

## 1.5  Conclusions

Future studies must clarify past extreme precipitation situations (including the socio-economic consequences) for every region or area so that we can

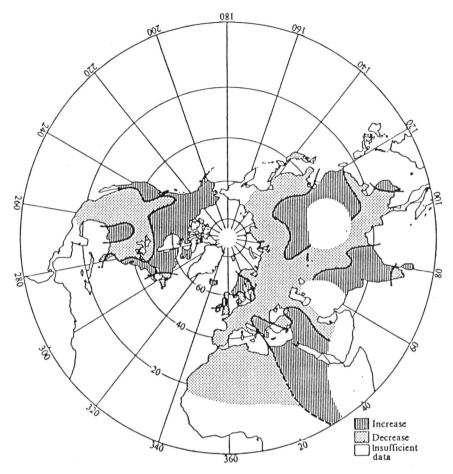

**Figure 1.5**  *Mean annual precipitation changes from cold to warm years (Wigley et al., 1980)*

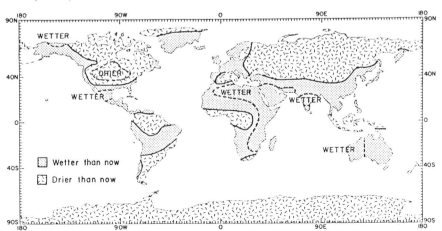

**Figure 1.6**  *Possible soil moisture distribution on a warmer earth, reconstructed from the paleoclimatic data of the Altithermal (Kellogg and Schware, 1981)*

supply political and administrative decision makers with relevant information. Our present climatological and meteorological knowledge gives no useful information on the change in precipitation conditions caused by global warming. Even a subtle change in general circulation can cause many harmful effects on a regional or smaller scale.

# References

JMA (1989) *Report on Recent Climatic Change in the World – Reviews and Outlook for the Future.*

Kellogg, W. W. and Schware, R. (1981) *Climate Change and Society – Consequences of Increasing Atmospheric Carbon Dioxide*, Westview Press, Boulder, CO.

Noda, A. and Tokioka, T. (1989) 'The effect of doubling the $CO_2$ concentration on convective and non-convective precipitation in a general circulation model coupled with simple mixed layer ocean model', *J. Meteor. Soc. Japan* (in press).

Theon, J. S. and Fugono, N. (eds) (1988) *Tropical Rainfall Measurements*, A.Deepak Publishing.

Wigley, T. M. L., Jones, P. D. and Kelly, P. M. (1980) 'Scenario for a warm, high-$CO_2$ world', *Nature*, **283**, 17–21.

WMO (1988a) *Concept of the Global Energy and Water Cycle Experiment*, WCRP-5, WMO/TD-No.215.

WMO (1988b) *The Global Precipitation Climatology Project*, Second Session of the International Working Group on Data Management, WCRP-6, WMO/TD-No.221.

# 2 PREDICTING TEMPORAL AND VOLUMETRIC CHANGES IN RUNOFF REGIMES UNDER CLIMATE-WARMING SCENARIOS

James M. Byrne, Doug Schaffer and Rod B. McNaughton
*University of Lethbridge, Alberta, Canada*

## 2.1 Introduction

The potential for global climate warming due to pollution from anthropogenic sources has raised serious concerns among water resource scientists and managers. Global circulation models (GCMs) have predicted there will be an increase in the earth's mean temperature of from 2°C to 5°C over the next 50–100 years. Most water resource planning decisions incorporate time frames of at least that length. Therefore planners today are making decisions in a void regarding both the quantity and temporal distribution of the future water supply.

Waggoner (1988) summarized the views of a number of scientists with the following recommendation:

> Scientists investigating climate change should make a special effort to improve predictions on the scales of time and space most relevant to the management of water resources, i.e., the scales of decades and the scales of large hydrologic surface and groundwater basins, setting bounds on the likely changes of averages, extremes, interannual variability and the rate of changes.

The philosophy here is in agreement with Waggoner. However, application of this philosophy is difficult due to a lack of specific knowledge on regional warming scenarios, and the secondary changes to climate parameters other than temperature. GCMs vary in climate parameter utilization and parameter interaction formulations; and typically have widely spaced resolution over the surface of the globe. For example, the model used at the US National Center for Atmospheric Research in Boulder, Colorado, uses a surface grid of 4.5° latitude and 7° of longitude (Schneider, 1984). This type of spacing provides very little regional detail; however, the limitations of the best available computers do not allow GCMs to operate with finer grid systems. This leaves most researchers and planners in somewhat of a dilemma. Typical efforts by these people are focused on areas of a much smaller scale than the GCM grid spacings; and in many cases, these areas exhibit diverse climate characteristics.

Much of the streamflow supply for western North America originates as snowmelt runoff from the Rocky Mountains. The Water Resources Institute is currently studying the potential net decline in water supply under climate warming for river basins with headwaters in the Rocky Mountains. The net decline will be the sum of several impacts: increased demand from irrigated areas due to higher temperatures; lower streamflow supplies because of changes to the snowmelt regime; and increased water utilization by natural vegetation. This chapter presents an assessment of potential changes to the precipitation regime for the eastern slopes of the Canadian Rocky Mountains, and the impacts these changes may have on streamflow quantities and temporal distributions.

## 2.2 Physiographic, climatic and hydrologic characteristics

The location of the region under study is presented in Figure 2.1. Figure 2.2 is an enlarged view of the study region, and includes the locations of the

**Figure 2.1**   *Study area location, Canada*

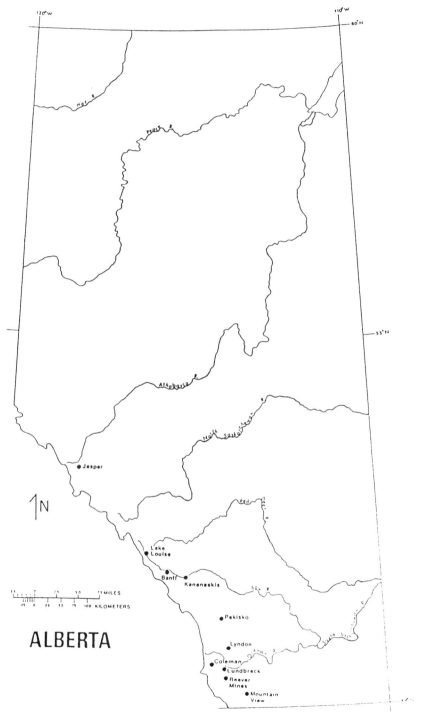

**Figure 2.2** *Climate stations and river basins under study*

climate stations and the river basins that are of interest to this study. The zone includes much of the area between 49° and 55° N and 110–120° W. The area exhibits a climatic and biologic diversity from west to east, with cool mountain climates in the west grading into semi-arid and arid plains in the east. Most of the region suffers from an average moisture deficit for grain crops of about 5 cm in the west to 30 cm in the east (*Atlas of Alberta*, 1969). Monthly precipitation peaks in May and June for the plains areas. Precipitation in summer is usually due to convective storms; hence the spatial distribution is erratic.

There are five river basins that drain the area. The southerly four, the Oldman, Bow, Red Deer and North Saskatchewan rivers, are part of the Saskatchewan river basin that runs east to Hudson Bay. The Athabasca river is part of the MacKenzie river drainage that runs to the Arctic ocean. The four southern rivers, particularly the Oldman and Bow basins, provide the water supply for most of the irrigated land in Canada. Much of the study focus is on these two basins, since the intensive irrigation developments result in these being at greater risk of suffering negative impacts due to climate warming. Salient hydrologic, climatic and water utilization data for these basins are summarized in Table 2.1.

**Table 2.1**   Important characteristics of the Oldman and Bow River basins.

| Parameter | Oldman River Bain | Bow River Basin |
|---|---|---|
| Drainage area | 15,761.4 km$^2$ | 18,119.6 km$^2$ |
| Irrigated areas | 253,000. ha | 231,000. ha |
| Population | | |
| Average Runoff | 3.5 × 10$^9$ m$^3$ | 3.2 × 10$^9$ m$^3$ |
| Average Irrigation Diversions | 1.4 × 10$^9$ m$^3$ | 1.3 × 10$^9$ m$^3$ |
| High runoff year | 6.2 × 10$^9$ m$^3$ | 6.9 × 10$^9$ m$^3$ |
| Low runoff year | 1.9 × 10$^9$ m$^3$ | 1.8 × 10$^9$ m$^3$ |
| Soil Moisture deficit (west to east) | 0. − 25 cm. | 0. − 25 cm. |
| All runoff values are estimates of natural conditions. | | |

## 2.3 Review and methodology

Since the 1970s, concerns for the effects of climatic warming on water demands and supplies have resulted in a number of studies of varied complexity. These include Schneider and Temkin (1977), Stockton and

Boggess (1979), Coutant (1981), Revelle and Waggoner (1982), Glieck (1987), Cohen (1987), Byrne (1990) and many others. The earliest of these studies either depended on qualitative judgements of the potential effects or utilized straightforward temperature, precipitation and runoff relationships. Several of them (Stockton and Boggess; Revelle and Waggoner; Byrne) utilized data originally developed by Langbein *et al.* (1949). Langbein presented average runoff values for the arid and semi-arid regions of the western USA as a function of the average annual temperature and precipitation. The relative changes in runoff that might be expected under the given climate scenarios were estimated with these relationships. Earlier studies carried out by Byrne (1989) fit a regression equation to the data with runoff depth $R$ as a function of the mean annual temperature $T$, and precipitation $P$. The $R^2$ value for this relationship is 0.94, and the standard error for $R$ estimate is 26 mm, or about 20% for the mountain climate stations that were assessed. The relationship is

$$R = -66.8 - 14.2T + 0.544P \tag{2.1}$$

This approach has a limited functionality. The initial data were developed from long-term averages for a wide geographic area. Application to specific locations is somewhat suspect, and there would be even less confidence in applying the equation to annual extremes in an attempt to define any level of variability and associated risk of runoff shortfall or flooding. In order to develop estimates of variability and risk level, one must adopt a modelling technique with a stronger theoretical basis. This type of development usually implies a greater complexity in terms of parameter representation, and the adoption of a temporal or time-series model that incorporates recorded or synthetic variability from season to season and year to year. This study presents two approaches. The first assessment utilizes weekly and monthly data to develop complementary precipitation and temperature distributions. The second employs a climatological model with a daily time step to simulate the effects of several climatic warming scenarios on critical seasonal hydrologic conditions.

## 2.4 Weekly/monthly analysis

This part of the study focused on assessing whether there are definite relationships between temperature and precipitation data for a mountain station, and if changes in the temperature regime affect variations in the precipitation amounts or temporal distribution(s). Figure 2.3 is a sample of the type of distribution that has been developed. The data for Figure 2.3 are from Jasper, a station in the northern part of the study area. The vertical bars based on the $x$-axis represent the percentage precipitation distribution as a function of mean temperature. This distribution was developed from mean weekly temperature and sum weekly precipitation

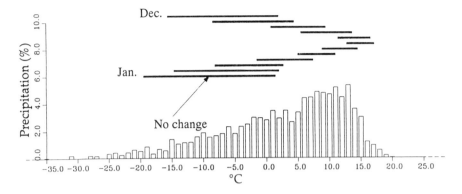

**Figure 2.3**   *Precipitation and temperature data for Jasper – no warming*

data for a period of record of 62 years. The distribution is distinctly skewed, with a peak in the range of 10–15°C. For Jasper, about 30% of the precipitation occurs as snow (temperatures below freezing), and 70% as rainfall. This breakdown is quite important, since the runoff regime for the rivers under study is a snowmelt regime. The much higher amounts of rainfall that occur contribute only a small amount to the runoff totals. This is particularly true in dry years, when the soil moisture storage deficit is usually capable of absorbing any rain that occurs.

The horizontal bars in Figure 2.3 represent the range of the mean monthly temperature for Jasper, 19 out of 20 years. These bars serve to illustrate several conceptual points. First, the warmer the winter months (November through March), the greater the relative amount of precipitation. Second, cooler summer months (July–August) receive greater precipitation. Third, the colder periods of the year demonstrate a much greater temperature variability than the warm periods.

These observations would lead to a prelimimary conclusion that climatic warming would result in Jasper having warmer, wetter winters and warmer, drier summers. However, the impacts of the wetter winters could be to cause runoff volumes either to increase or to decline, depending on the change in the precipitation phase – greater snowfall will cause an increase in runoff; lower snowfall totals will lead to a decline. One obvious conclusion is the snowfall season will definitely be shorter, and the snowmelt runoff will occur earlier in the spring. This is not expected, and is supported by conclusions in other work (Gleick, 1987).

A simple way to estimate the changes that may occur is to shift the temperature distributions upwards to the predicted level of climate warming. This may be an adequate approach if one is interested in the changes in mean values. However, the interest here is to define the changes in variability as well. A general conclusion of many of the global climate model programmes has been that climate variability will increase with global warming. The increased variability, as discussed in Hansen *et al.*(1988), will be manifest in greater precipitation and drought extremes.

This is not unexpected – global warming will result in a lower temperature gradient from the equator to the polar regions. This gradient is the energy source for the global circulation. Weakening of the gradient should result in a weakened circulation that will conform to the seasonal controls exerted by oceans and landmasses.

Overall, the temperature variability for the study area will probably decline. The warmer the average temperatures, the lower the extremes that have been experienced. Figure 2.4 presents the relationship of average monthly temperature to monthly standard deviation for ten climate stations. The figure indicates that the warmer the monthly temperature, the lower the variation in the temperature. The standard deviation $\sigma$ of monthly mean temperature declines exponentially with increasing mean temperatures $T_m$. The equation that describes the function in Figure 2.4 is

$$\sigma = 2.643 \exp(-0.0513T_m) \tag{2.2}$$

This equation has an $R^2 = 0.862$. The argument here is the change in temperature variability with climatic warming will probably behave according to the function in Figure 2.4. Therefore, the range in mean monthly temperatures, as illustrated in Figure 2.3, will not be applicable once a 'warming' of several degrees has been applied. Equation (2.2) is used to determine the extent of the monthly temperature distributions once the means were adjusted upwards.

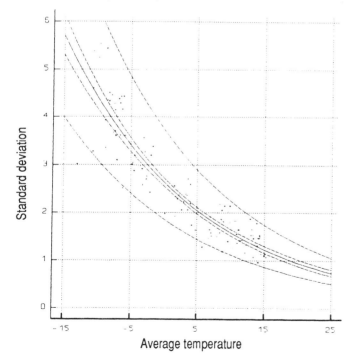

**Figure 2.4**  *Temperature variability versus average monthly temperature*

A second concern must be addressed prior to making any temperature adjustments. GCMs have generally predicted that the expected warming will be greater in winter periods than in summers (Environment Canada, 1988). An average 2°C annual warming may mean a warming of twice that magnitude in winter. A simplistic adjustment was applied to account for the seasonal variation in warming. Warming in December and January was assumed to be twice the values for June and July, with a linear decline in the multiplier from winter to summer.

The new monthly temperature regime for Jasper is presented in Figure 2.5. An annual warming of 1.5°C has been applied to the monthly means, with the seasonal adjustment as discussed above. There are several points that may be made with respect to the figure. The range of temperature variability declines significantly in the winter months. The result is that the upper temperature extremes for the coldest months are very close to those of the unadjusted data; but the mean and lower extremes for the adjusted data are significantly higher. Since potential water vapour storage in an airmass increases exponentially with increasing temperature, there is a potential for a greater amount of snowfall during the months of December, January, February and March. During those months, temperatures would remain close to (but very rarely be above) the freezing point. The potential for greater winter snowfall, and consequently increased runoff in the spring, is not unreasonable. A warmer Pacific to the west, where most of the region's precipitation now originates, may provide the additional moisture.

The concepts presented in Figure 2.5 argue that the summers would be warmer, drier, and slightly more stable. Again, this is not unexpected, and is in agreement with GCM predictions (Hansen *et al.*, 1988) and other similar studies (Gleick, 1987). The precipitation distribution declines very rapidly with temperatures over 13°C. The temperatures of the warmest months (July and August) have shifted upwards such that, in most years, the lower extremes would be equivalent to recorded average temperatures.

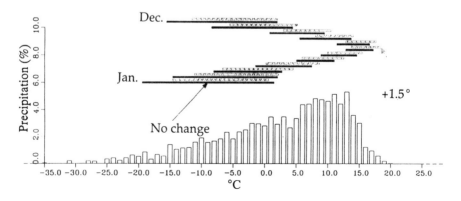

**Figure 2.5**   *Precipitation and temperature data for Jasper – 1.5°C average annual warming*

This puts the temperature distributions for these months well above the precipitation distribution peak at 13°C.

## 2.5 Daily assessment of climatic/hydrologic conditions

Between 70% and 90% of the annual volume of streamflow in the Bow and Oldman rivers is due to snowmelt runoff in the mountains and foothills. Therefore, the greatest potential for adversely affecting the water supply for the region would lie in changes to the winter precipitation regime. The intent of the daily modelling scheme is to simulate the water equivalent of the spring snowpack for climatic stations in the Rocky Mountains under various levels of climatic warming. Comparison of these simulations to a base case simulation with no warming impact (i.e. the recorded climatic situation) will provide some estimate of the hydrologic impacts that climate warming may have on the area. Two critical issues will be addressed: (1) the volumetric changes that might occur due to reduction in the size of the snowpack, and (2) changes in the temporal characteristics of runoff due to snowmelt occurring earlier in the year.

*2.5.1 Climatic model for determining spring snow water equivalent*

The climatic/hydrologic model utilized has been previously verified for application in the study area (Byrne, 1990, 1989). The model calculates the spring snow water equivalent (*SWE*) at a climate station as:

$$SWE = \sum_{i=\text{Nov.1}}^{ETDAY} [SNO_i + RAIN_i - V_{si}] \qquad (2.3)$$

where

$SNO_i$ and $RAIN_i$ = daily snowfall and rainfall,

$V_{si}$ = daily net vapour transfer to/from the snowpack.

*ETDAY* is the day the model determines the snowmelt period is over due to exhaustion of the snowpack by the snowmelt routine (Quick and Pipes, 1977). From *ETDAY* forward, the model begins to utilize the soil moisture functions in the water balance modelling. It is important to note that *ETDAY* is a 'hydrologic' date and will occur on different days for different stations, depending generally on the local station climate situation each spring.

Morton's (1978) evaporation/evapotranspiration model was utilized to estimate the net vapour transfer $V_{si}$ to/from the snowpack over winter.

Morton's model is based on the principle that regional temperature and humidity data reflect the availability of water for evaporation/sublimation. No other functional winter sublimation/condensation model is known to be available at this time. Morton's model utilizes temperature, humidity and sunshine data for a station to estimate the net moisture transfer to the atmosphere. Using this model presented a number of difficulties in terms of adjustments to the input parameters, primarily because the GCM warming scenarios do not provide sufficient detail on expected changes to climate parameters other than temperature. The following assumptions were applied in adjusting the climate data to reflect a warmer regional climate.

Sunshine data will not be directly affected by climate warming, but could certainly be altered by changes in the seasonal and annual cloudiness. There is no information available from either GCMs or other sources that provides any justification for adjustment of sunshine hours. Consequently, the unadjusted recorded data are utilized in the simulations for all scenarios. Earlier discussion indicated that temperatures are predicted to rise in the range of 2–5°C on an annual basis, with predicted warming in the winter of twice the annual value. An adjustment curve was used in the weekly–monthly analysis to reflect this, and the same method was applied here. A 1°C warming in June and July translated to a 2°C warming in December and January, with a linear rate of change between.

Adjustment of humidity and precipitation data to reflect climatic warming is tenuous, at best. The Rocky Mountains form a barrier between the maritime climate associated with the Pacific to the west and the continental conditions that dominate most of the central part of North America. A warmer Pacific will probably cause greater precipitation to the west, but the climate to the east will become drier under climate warming. The boundary between these two dominant climate controls shifts back and forth over the study area now, and will probably continue to do so under a warming situation. There may not be a change in the long-term mean humidity and precipitation, but there would be an increase in the variability. This is very difficult to assess quantitatively. The assumption here is that any attempt to impose a change with no supporting data would be counterproductive, and possibly misleading. Therefore, there were no changes made to the recorded precipitation. The humidity conditions, as reflected by the dewpoint temperatures, were subjected to a slight relative decrease in most scenarios. The actual dewpoints were increased in most cases by the same amount as the air temperatures. However, the saturation vapour pressure curve is non-linear with temperature, so that the shift results in there being slightly lower estimates of $V_{si}$ because of the difference between the saturation and actual vapour pressures.

## 2.5.2 Climate-warming scenarios

The climate model has been utilized to estimate the spring snow water equivalent and *ETDAY* for each year of record for the ten climate stations for the following warming scenarios:

1. No change;
2. Annual average warming of 1.5°C: 1°C in summer, 2°C in winter;
3. Annual average warming of 3.0°C: 2°C in summer, 4°C in winter.

## 2.5.3 Results and discussion

The results of the daily simulations are presented in Table 2.2. The data for Coleman are presented as representative of the winter trends for the climate stations. Coleman is located in the southern part of the study area (see Figure 2.2). The second, fifth, and eighth columns are estimates of the value of *ETDAY* each year for the three scenarios. (*ETDAY* was defined above as the day the model determines the snowmelt period is over due to exhaustion of the snowpack by the snowmelt routine (Quick and Pipes, 1977)). The occurrence of *ETDAY* is critical, since an earlier spring will result in an earlier runoff period. In many areas, this would have an adverse effect on irrigation developments if there was insufficient storage available to hold the runoff for use during the irrigation season.

The mean and standard deviation ($\sigma$) of the data columns is included at the bottom of Table 2.2. With increasing climate-warming scenarios, the mean occurrence of *ETDAY* shifts earlier in the year. Under scenario 1 (no charge), the mean date of occurrence for *ETDAY* is day 120 (30 April), and $\sigma = 14$ days. Assuming the data are normally distributed, then *ETDAY* has been occurring between 2 April and 28 May, 19 years out of 20 ($\pm 2\sigma$ is 95% of cases). However, under scenario 2 (annual warming of 1.5°C), the mean date for *ETDAY* is 22 April, and the range of dates is 22 March to 22 May for $\pm 2\sigma$. Under scenario 3 (annual warming of 3.0°C), the mean for *ETDAY* is 12 April, and the range for 19 out of 20 years would be 12 March to 14 May.

The changes in the *ETDAY* distribution under climate warming are quite severe. The natural flow conditions for rivers in the study area result in a very rapid drop in discharge from early to midsummer. The changes in *ETDAY* would result in a severe reduction in streamflow supplies at a time when the irrigation demand is increasing significantly. The result would be a temporal water supply problem that, in dry years, would be extreme. There would be some offsetting benefit from planting crops earlier in the year. The latter date for the *ETDAY* distribution moves from 28 May to 22 May for a 1.5°C warming, and to 14 May for a 3.0°C warming. The agricultural community could resort to earlier planting as one means to continue use of the earlier runoff water supply. However, this is

**Table 2.2** Spring conditions at Coleman, four climate scenarios

| Year | ETDAY1 | SWE1 | ET1 | ETDAY2 | SWE2 | ET2 | ETDAY3 | SWE3 | ET3 |
|---|---|---|---|---|---|---|---|---|---|
| 1915 | 107 | 183 | 58 | 93 | 138 | 42 | 89 | 143 | 36 |
| 1916 | 110 | 169 | 55 | 99 | 157 | 47 | 96 | 165 | 39 |
| 1917 | 141 | 182 | 75 | 118 | 161 | 57 | 116 | 142 | 56 |
| 1918 | 114 | 417 | 51 | 100 | 423 | 45 | 98 | 422 | 44 |
| 1919 | 113 | 150 | 58 | 109 | 157 | 51 | 108 | 162 | 46 |
| 1920 | 129 | 313 | 63 | 126 | 316 | 61 | 123 | 315 | 56 |
| 1921 | 120 | 340 | 65 | 119 | 296 | 61 | 108 | 285 | 46 |
| 1922 | 129 | 216 | 68 | 128 | 218 | 66 | 113 | 195 | 53 |
| 1923 | 120 | 201 | 57 | 119 | 206 | 53 | 104 | 218 | 41 |
| 1924 | 131 | 100 | 69 | 116 | 110 | 53 | 112 | 116 | 47 |
| 1925 | 108 | 181 | 52 | 102 | 184 | 47 | 99 | 188 | 43 |
| 1926 | 121 | 33 | 64 | 120 | 41 | 49 | 119 | 48 | 39 |
| 1927 | 132 | 171 | 61 | 122 | 160 | 56 | 107 | 135 | 49 |
| 1928 | 132 | 209 | 76 | 131 | 210 | 75 | 96 | 217 | 39 |
| 1929 | 129 | 168 | 58 | 125 | 155 | 56 | 124 | 149 | 52 |
| 1930 | 110 | 164 | 58 | 105 | 171 | 50 | 104 | 172 | 49 |
| 1931 | 115 | 115 | 63 | 113 | 128 | 50 | 109 | 133 | 44 |
| 1932 | 111 | 94 | 61 | 109 | 94 | 60 | 98 | 107 | 47 |
| 1933 | 130 | 137 | 67 | 128 | 138 | 64 | 122 | 140 | 60 |
| 1934 | 112 | 257 | 66 | 110 | 266 | 57 | 108 | 264 | 48 |
| 1935 | 129 | 342 | 61 | 127 | 350 | 53 | 126 | 342 | 45 |
| 1936 | 123 | 122 | 65 | 122 | 128 | 59 | 122 | 132 | 55 |
| 1937 | 117 | 212 | 55 | 114 | 217 | 49 | 112 | 216 | 48 |
| 1938 | 118 | 235 | 59 | 118 | 241 | 50 | 88 | 224 | 27 |
| 1939 | 99 | 107 | 38 | 99 | 110 | 32 | 98 | 110 | 30 |
| 1940 | 123 | 130 | 67 | 122 | 136 | 60 | 92 | 107 | 30 |
| 1941 | 101 | 108 | 38 | 97 | 116 | 31 | 96 | 123 | 24 |
| 1942 | 120 | 92 | 48 | 109 | 98 | 34 | 106 | 102 | 26 |
| 1943 | 111 | 185 | 49 | 106 | 192 | 41 | 103 | 194 | 40 |
| 1944 | 113 | 20 | 44 | 111 | 26 | 38 | 110 | 19 | 32 |
| 1945 | 139 | 204 | 56 | 138 | 215 | 45 | 127 | 183 | 40 |
| 1946 | 92 | 142 | 46 | 86 | 148 | 33 | 71 | 150 | 18 |
| 1947 | 105 | 247 | 59 | 97 | 260 | 47 | 96 | 269 | 37 |
| 1948 | 128 | 142 | 70 | 127 | 148 | 64 | 126 | 153 | 58 |
| 1949 | 105 | 85 | 47 | 100 | 69 | 46 | 98 | 68 | 46 |
| 1950 | 144 | 283 | 66 | 119 | 266 | 54 | 101 | 256 | 39 |
| 1951 | 134 | 317 | 94 | 133 | 293 | 88 | 103 | 265 | 49 |
| 1952 | 109 | 133 | 59 | 106 | 133 | 58 | 105 | 137 | 54 |
| 1953 | 132 | 287 | 59 | 131 | 288 | 58 | 130 | 214 | 39 |
| 1954 | 146 | 537 | 73 | 145 | 536 | 69 | 144 | 538 | 67 |
| 1955 | 116 | 120 | 53 | 115 | 121 | 49 | 111 | 121 | 41 |
| 1956 | 147 | 203 | 109 | 146 | 168 | 92 | 97 | 155 | 42 |
| 1957 | 126 | 144 | 66 | 124 | 149 | 59 | 92 | 164 | 36 |
| 1958 | 116 | 122 | 47 | 110 | 115 | 38 | 108 | 124 | 30 |
| 1959 | 137 | 282 | 85 | 135 | 251 | 72 | 83 | 218 | 34 |
| 1960 | 141 | 181 | 76 | 140 | 188 | 69 | 98 | 205 | 40 |
| 1961 | 94 | 163 | 39 | 91 | 163 | 32 | 87 | 153 | 25 |
| 1962 | 109 | 146 | 59 | 103 | 155 | 50 | 97 | 154 | 41 |
| 1963 | 98 | 126 | 41 | 95 | 130 | 30 | 57 | 121 | 13 |
| 1964 | 118 | 191 | 59 | 111 | 194 | 43 | 110 | 191 | 35 |
| 1965 | 119 | 287 | 67 | 115 | 293 | 61 | 113 | 296 | 55 |
| 1966 | 139 | 202 | 87 | 102 | 181 | 49 | 89 | 178 | 40 |
| 1967 | 144 | 260 | 82 | 143 | 268 | 74 | 112 | 242 | 47 |
| 1968 | 133 | 119 | 64 | 115 | 102 | 46 | 72 | 84 | 25 |
| 1969 | 109 | 168 | 61 | 97 | 181 | 46 | 95 | 176 | 44 |
| 1970 | 139 | 101 | 91 | 98 | 94 | 45 | 93 | 95 | 43 |
| 1971 | 114 | 309 | 63 | 113 | 307 | 60 | 107 | 295 | 51 |
| 1972 | 130 | 453 | 76 | 116 | 443 | 59 | 115 | 428 | 54 |
| 1973 | 118 | 90 | 62 | 75 | 76 | 35 | 72 | 82 | 29 |
| 1974 | 113 | 223 | 72 | 108 | 230 | 64 | 106 | 200 | 58 |
| 1975 | 128 | 157 | 65 | 126 | 148 | 63 | 123 | 146 | 58 |
| 1976 | 113 | 168 | 64 | 101 | 181 | 46 | 97 | 190 | 37 |
| 1977 | 115 | 34 | 58 | 113 | 28 | 48 | 101 | 40 | 32 |
| 1978 | 98 | 114 | 51 | 95 | 116 | 49 | 94 | 117 | 48 |
| 1979 | 119 | 193 | 57 | 117 | 192 | 58 | 116 | 196 | 54 |
| 1980 | 116 | 148 | 70 | 106 | 165 | 52 | 92 | 169 | 40 |
| 1981 | 123 | 198 | 67 | 81 | 215 | 32 | 67 | 221 | 20 |
| 1982 | 130 | 172 | 68 | 122 | 174 | 58 | 120 | 179 | 53 |
| 1983 | 123 | 126 | 70 | 105 | 123 | 49 | 61 | 113 | 18 |
| 1984 | 101 | 118 | 48 | 98 | 93 | 42 | 97 | 87 | 30 |
| 1985 | 109 | 53 | 59 | 92 | 57 | 46 | 89 | 52 | 42 |
| 1986 | 80 | 167 | 33 | 79 | 165 | 30 | 78 | 170 | 23 |
| 1987 | 112 | 112 | 68 | 110 | 104 | 59 | 109 | 113 | 47 |
| Average | 120 | 183 | 62 | 112 | 182 | 52 | 102 | 178 | 41 |
| Std dev. | 14 | 93 | 13 | 15 | 92 | 12 | 16 | 89 | 11 |

somewhat simplistic, and may be a difficult strategy to adopt, given the general inability to differentiate between natural variation and the effects of 'greenhouse warming'.

The above discussion does not apply to the stations in the northern part of the study area. The data for Jasper illustrate this. Under scenario 1 (no change), the mean date of occurrence for *ETDAY* is day 114 (24 April), and $\sigma = 11$ days. Assuming the data are normally distributed, then *ETDAY* has been occurring between 2 April and 16 May, 19 years out of 20 ($\pm 2\sigma$ is 95% of cases). However, under scenario 2 (annual warming of 1.5°C), the mean date for *ETDAY* is 16 April, and the range of dates is 19 March to 14 May for 19 out of 20 years. Under scenario 3 (annual warming of 3.0°C), the mean for *ETDAY* is 7 April, and the range 19 out of 20 years would be from 2 March to 13 May.

The changes in the *ETDAY* distribution under climate warming for Jasper are also quite severe. A particularly important point to note is there would be minimal benefit from planting crops earlier in the year. The latter date for the *ETDAY* distribution moves very little, from 16 May to 13 May. This implies there would be very little opportunity to plant frost-sensitive crops earlier in the year, so as to take advantage of the earlier runoff.

The discussion of simulated changes to *ETDAY* is subject to a caveat. Both stations are mountain stations, therefore the *ETDAY* migration under climate warming may be very different from what may occur on the plains to the east, where all the agricultural production takes place. This aspect is currently under study, but no definitive results have yet been generated. However, the plains will always experience an earlier spring than the mountains. The question is whether the shift in *ETDAY* for the plains will be proportionate to that experienced in the mountains or will be greater. Conventional climatology would argue that the shift in *ETDAY* for the plains would not be less than that experienced in mountains.

Columns three, six and nine of Table 2.2 are the estimates of the spring snow water equivalent for scenarios one, two and three at Coleman. The model predicts very little change in the value of *SWE* with climate warming. This is somewhat surprising, since less precipitation is input to the model due to the shortening of the accumulation season with each warming scenario. The net water transfer $V_{si}$ from the snowpack over winter is presented in columns four, seven and ten of Table 2.2. The trends in the means and standard deviations indicate that $V_{si}$ declines with warming. This is opposite to the expected trend. This is partly due to the assumptions of the Morton vapour transfer model. Equal increases in the dewpoint and daily temperature will induce a slight downward bias in the $V_{si}$ estimates. However, this will not be a significant error unless there was a major change in the regional humidity regime. Further information would be required from GCMs to determine if such a change is likely to take place in the study area.

## 2.6 Conclusions

The major empirical observations of this chapter are:

1. Climate warming in winter increases the probability of heavier winter snowfall, assuming the moisture is available. This may be the case for the study area due to the proximity of a warmer Pacific.
2. The summer conditions have an increased probability of lower precipitation totals.
3. There is no evidence to suggest that spring snow water equivalent values will seriously decline with significant climate warming.
4. The winter season will shorten significantly, resulting in earlier spring runoff. The availability of streamflow water supplies under natural runoff conditions will decline in the summer months.

These results were generated from two different modelling strategies. First, the distribution of precipitation by temperature was developed based on monthly averages. The existing temperature distribution was then shifted upwards, and adjusted for the likely decrease in variability at higher temperatures. The new distribution does not provide evidence of any significant decrease in winter snowfall. Thus, current levels of spring runoff volumes should be maintained. Second, a climatic/hydrologic model was developed to predict spring SWE under two climatic warming scenarios: an average annual warming of (1) 1.5°C and (2) 3.0°C. This modelling confirmed the earlier findings regarding winter precipitation: climatic warming would not result in a decrease in spring runoff. A shortening of the winter season is likely, however, suggesting that runoff will occur earlier in the spring. This may cause a relative deficit of streamflow in summer months.

The implications of these findings for the management of water resources in watersheds to the lee of the Rocky Mountains is clear. These watersheds contain about 25% of Canada's irrigated lands, and any reduction in summer natural streamflow will have a negative impact on agricultural production if there is insufficient storage to adequately manage the water supply. It is unlikely that the earlier spring runoff can be captured through earlier planting of crops. The temperature distribution suggests that although spring warming will occur earlier on average, the date of the last serious frost risk may not shift to the same degree. This is particularly true in the more northerly parts of the study area.

## 2.7 Recommendation

This work is predicated on a very weak knowledge of the regional warming that may occur under general global warming. There is a strong need for much better regional data, and the only way to achieve this is to improve GCMs. The scientific community should offer support wherever it may be required to enable GCM programmes to obtain the resources necessary to improve the resolution and parameterization of global models. There is no other source for acquiring detailed regional warming scenarios, and until these are available, scientists and planners working on a regional scale will be operating in a void.

# References

*Atlas of Alberta* (1969) University of Alberta Press, Edmonton, Canada.

Byrne, J. M. (1989) 'Three phase runoff model for small prairie rivers: I. Frozen soil phase assessment', *Canadian Water Resources Journal*, **14**, 17–28.

Byrne, J. M. (1990) 'Assessing potential climate change impacts on water supply and demand in southern Alberta', *Canadian Water Resources Journal*, in press.

Cohen, S. (1987) 'Projected increases in municipal water use in the Great Lakes due to $CO_2$-induced climatic change', *Water Resources Bulletin*, **23**, 91–101.

Coutant, C. C. (1981) 'Foreseeable effects of $CO_2$-induced climatic change: freshwater concerns', *Environmental Conservation*, **8**, 285–297.

Environment Canada (1988) Conference Statement, Toronto: *The Changing Atmosphere – Implications for Global Security*, hosted by Environment Canada, 27–30 June.

Gleick, P. H. (1987) 'The development and testing of a water balance model for climate impact assessment: modeling the Sacramento Basin', *Water Resources Research*, **23**, 1049–1061.

Hansen, J., Rind, D., Delgenio, A., Lacis, A., Lebedeff, S., Prather, M., Ruedy, R. and Karl, T. (1988) 'Regional greenhouse climate effects', *Proceedings of the Second North American Conference on Preparing for Climate Change*, Washington, DC, 6–8 December.

Langbein, W. B. *et al.* (1949) *Annual runoff in the United States*, US Geological Survey Circular 5, US Department of the Interior, Washington, DC.

Morton, F. I. (1978) 'Estimating evapotranspiration from potential evaporation: practicality of an iconoclastic approach', *Journal of Hydrology*, **38**, 1–32.

Pipes, A. and Quick, M. C. (1977) *UBC watershed model users guide*, Dept of Civil Engineering, University of British Columbia.

Revelle, R. R. and Waggoner, P. E. (1983) 'Effects of a carbon dioxide induced climate change on water supplies in the western United States', in *Changing Climate: Report of the USNRC Carbon Dioxide Assessment Committee*.

Schneider, S. H. (1984) 'Climate modeling', *Scientific American*, **256**, 72–80.

Schneider, S. H. and Temkin, R. L. (1977) 'Water supply and future climate', in *Studies in Geophysics, Climate, Climate Change and Water Supply*, National Academy of Science, Washington, DC.

Stockton, C. W. and Boggess, W. R. (1979) *Geohydrological implications of climate change on water resource development*, US Army Coastal Engineering Research Center, Fort Belvoir, Virginia.

Waggoner, P. E. (1988). From *Climate Changes and U. S. Water Resources*: A brief summary and the recommendations by Paul E. Waggoner (ed.), Proceedings of the Second North American Conference on Preparing for Climate Change, Washington, DC, 6–8 December.

# 3 EFFECT OF PRECIPITATION VARIABILITY ON RECHARGE IN UNCONFINED AQUIFERS

Abdulaziz S. Al-Turbak and Fouad F. Al-Muttair
*King Saud University, Saudi Arabia*

## 3.1 Introduction

Groundwater constitutes the most important natural water resource in Saudi Arabia. It exists in two different types of formation. The first is deep (mostly confined) aquifers throughout the eastern two-thirds of the country. These aquifers contain large amounts of water and direct recharge from precipitation to them is not significant. There are seven major aquifers of this type as well as many secondary confined ones (*Water Atlas of Saudi Arabia*, 1984).

The second type of aquifer is shallow unconfined (alluvial). These are scattered throughout the country and found under wadis and between escarpments. They are normally unconsolidated and of limited thickness, rarely exceeding 100 m (Rizaiza *et al.*, 1989). Alluvial aquifers have been sustaining agricultural development in Saudi Arabia for hundreds of years before drilling of wells in the large confined aquifers, which started only in the last few decades. They supply substantial amounts of water to agricultural areas located on many wadis.

To properly manage the water resources in confined aquifers, the Ministry of Agriculture and Water (MAW) has built about 200 dams (Al-Muttair *et al.*, 1986). These small to medium-sized dams were built with the purpose of storing floodwater flowing in wadis in reservoirs located just upstream of agricultural areas. In many cases, they also serve as flood-control structures preventing excessive flooding of villages and farms. It has been shown by many investigators that these dams will provide substantial amounts of recharged water to the aquifers downstream. (Al-Dalooj *et al.*, 1983; Al-Turbak and Al-Muttair, 1989).

The major objective of this chapter is to show the effect of variability of precipitation in a three-year period on recharge in parts of two unconfined aquifers located in central Saudi Arabia, just downstream from two recharge dams. Data on rainfall were taken from two meteorological stations installed close to these dams. Few observation wells exist in each locality, and the water level data from these wells are used in the study. The observation wells are located just downstream from reservoirs that are used for recharge purposes. The water levels and volumes existing in the reservoirs for the three-year period are also available. Data on rainfall, water levels in the observation wells and volumes of water in the reservoirs will be used to analyse the effect of precipitation and its variability on recharge in the two shallow aquifers under consideration.

## 3.2 Study areas

The first catchment of Malham Dam is located approximately between 25° 01' N–25° 11' N latitude and 45° 58' E–46° 14' E longitude. It constitutes an area of about 289 km². Malham Dam itself is located 75 km north-west of Riyadh at the intersection of 25° 08' N latitude and 46° 14' E longitude. It is of a rockfill type with length 100 m, height 5 m, storage capacity of about 0.5 MCM and was constructed in 1970. Figure 3.1 shows the layout of the dam, wadi channel and the four observation wells (M2 upstream, and M1, S32 and S33 downstream) available at the site. The town of Malham is about 2 km downstream of the dam.

The second catchment of Al-Amalih Dam is located between 25° 32' N–25° 36' N latitude and 45° 32' E–45° 36' E longitude. The catchment area is about 21.6 km². Al-Amalih Dam is located at about 25° 35' N latitude and 45° 35' E longitude. The town of Hawtat Sudair is about 2 km downstream from the dam and about 170 km north of Riyadh. Al-Amalih Dam is of a concrete type with a length of 500 m, height 8 m, a storage capacity of about 1.0 MCM, and was completed in 1982. Figure 3.2 shows the dam, wadi channel and the four observation wells S4, S1, S2 and S3.

At each location (Malham and Al-Amalih) there is a meteorological station which includes a wind recorder, an actinograph, a Class A pan, a recording rain gauge, a totalizing rain gauge, a recording evaporation

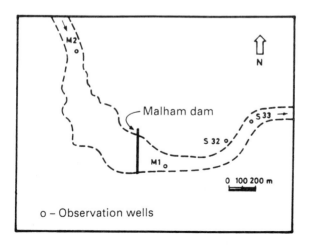

**Figure 3.1**   *Location of dam, wadi channel and observation wells, Malham*

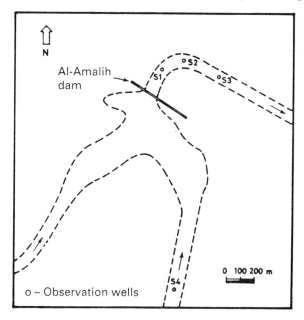

**Figure 3.2** *Location of dam, wadi channel and observation wells, Al-Amalih*

balance and a shelter containing a thermohygrograph, extreme thermo-meters and a Piche tube.

## 3.3 Data analysis

Tables 3.1–3.3 show precipitation, inflow to reservoir, and volumes of water infiltrated from the reservoirs ($R_v$) at Malham site during the 1985/1986, 1986/1987 and 1987/1988 seasons, respectively. The corresponding results for the Al-Amalih site are shown in Tables 3.4–3.6. The monthly precipitation amounts (mm) shown in these tables were obtained from the records of the two meteorological stations. The inflows were calculated for each reservoir using data from water level recorders installed on the reservoirs and volume–stage relationships that were established during repeated surveys of the two reservoirs (one survey before each season). The values shown in the last column of each table ($R_v$) represent the monthly volumes of water infiltrating through the bottom of the reservoir (monthly volumes of recharged water). These were calculated using a water budget model of the reservoir (Al-Turbak and Al-Muttair, 1989).

Figures 3.3 and 3.4 show the water levels in the two observation wells M1 and S2 from 15 January to 15 July during the three seasons 1985/1986, 1986/1987 and 1987/1988. They were obtained from data collected by water-level recorders installed on all observation wells and are shown here only as examples. The same trends exist for the other wells.

**Table 3.1**   Precipitation, inflow and recharge volume
(R$_v$) at Malham (1985/1986 season)

| Month | Precipitation (mm) | Inflow (m³) | R$_v$ (m³) |
|---|---|---|---|
| Nov. 1985 | 2.45 | – | – |
| Dec. 1985 | 19.82 | 30 984 | 16 835 |
| Jan. 1986 | 17.00 | 81 601 | 21 862 |
| Feb. 1986 | 7.44 | – | 59 832 |
| March 1986 | 44.80 | 702 656 | 598 883 |
| April 1986 | 89.95 | 1 079 465 | 722 545 |
| May 1986 | 3.50 | – | 177 876 |
| June 1986 | – | – | 58 473 |
| July 1986 | – | – | 17 710 |
| Aug. 1986 | – | – | 7,996 |
| Sep. 1986 | – | – | 4 004 |
| Total | 184.96 | 1 894 706 | 1 686 016 |

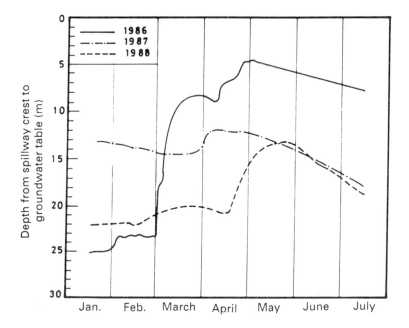

**Figure 3.3**   *Water levels in observation well M1, Malham*

**Table 3.2**   Precipitation, inflow and recharge volume
($R_v$) at Malham (1986/1987 season)

| Month | Precipitation (mm) | Inflow ($m^3$) | $R_v$ ($m^3$) |
|-------|-------------------|----------------|---------------|
| Dec. 1986 | 14.05 | – | – |
| Jan. 1987 | 0.20 | – | – |
| Feb. 1987 | 1.05 | – | – |
| March 1987 | 49.40 | 273 108 | 107 160 |
| April 1987 | 1.25 | – | 96 812 |
| May 1987 | – | – | 13 884 |
| June 1987 | – | – | 5 650 |
|  |  |  | 223 506 |
| Total | 65.95 | 273 108 | |

**Table 3.3**   Precipitation, inflow and recharge volume
($R_v$) at Malham (1987/1988 season)

| Month | Precipitation (mm) | Inflow ($m^3$) | $R_v$ ($m^3$) |
|-------|-------------------|----------------|---------------|
| Dec. 1987 | 0.60 | – | – |
| Jan. 1988 | 14.25 | – | – |
| Feb. 1988 | 23.90 | 40 709 | 22 824 |
| March 1988 | 8.70 | 2 517 | 10 902 |
| April 1988 | 35.85 | 601 273 | 355 600 |
| May 1988 | – | – | 151 919 |
| June 1988 | – | – | 19 089 |
| July 1988 | – | – | 1 623 |
| Aug. 1988 | – | – | 1 243 |
| Total | 83.30 | 644 499 | 563 200 |

The highest levels of the water table in the three observation wells existing downstream of Malham Dam (during the three seasons of the study) are shown in Table 3.7 with dates on which these levels were observed. Table 3.8 gives the same data for the three downstream wells at the Al-Amalih site. The values in Tables 3.7 and 3.8 represent the distances between the water table and the spillway crest. Therefore, the smaller the value shown, the higher the water level in the particular observation well.

**Table 3.4** Precipitation, inflow and recharge volume
($R_v$) at Al-Amalih (1985/1986 season)

| Month | Precipitation (mm) | Inflow ($m^3$) | $R_v$ ($m^3$) |
|---|---|---|---|
| Nov. 1985 | 19.60 | 329 189 | 268 660 |
| Dec. 1985 | 14.35 | 6 270 | 25 113 |
| Jan. 1986 | 16.55 | 54 795 | 18 623 |
| Feb. 1986 | 16.60 | 43 668 | 36 798 |
| March 1986 | 41.85 | 327 862 | 335 139 |
| April 1986 | 57.95 | 327 619 | 305 325 |
| May 1986 | 8.45 | – | 20 874 |
| June 1986 | – | – | 4 329 |
| July 1986 | – | – | 1 719 |
| Aug. 1986 | – | – | 181 |
| Sep. 1986 | – | – | 83 |
| Total | 175.45 | 1 089 403 | 1 016 844 |

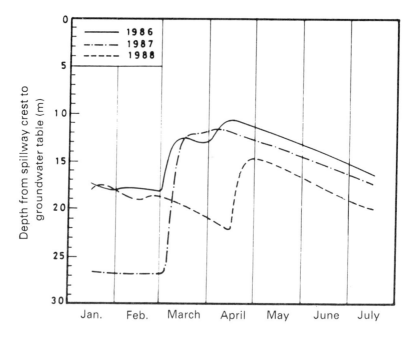

**Figure 3.4** *Water levels in observation well S2, Al-Amalih*

**Table 3.5** Precipitation, inflow and recharge volume ($R_v$) at Al-Amalih (1986/1987 season)

| Month | Precipitation (mm) | Inflow (m³) | $R_v$ (m³) |
|---|---|---|---|
| Nov. 1986 | 17.45 | 3 679 | 408 |
| Dec. 1986 | 6.60 | – | 1 846 |
| Jan. 1987 | 0.70 | – | 393 |
| Feb. 1987 | 8.80 | 286 | 116 |
| March 1987 | 66.65 | 720 151 | 577 383 |
| April 1987 | 0.35 | – | 95 147 |
| May 1987 | – | – | 7 462 |
| June 1987 | – | – | 2 015 |
| July 1987 | – | – | 570 |
| Total | 100.55 | 724 116 | 685 340 |

**Table 3.6** Precipitation, inflow and recharge volume ($R_v$) at Al-Amalih (1987/1988 season)

| Month | Precipitation (mm) | Inflow (m³) | $R_v$ (m³) |
|---|---|---|---|
| Oct. 1987 | 0.25 | – | – |
| Nov. 1987 | – | – | – |
| Dec. 1987 | 1.60 | – | – |
| Jan. 1988 | 21.85 | 125 929 | 123 662 |
| Feb. 1988 | 36.25 | 20 975 | 17 984 |
| March 1988 | 1.30 | – | 1 237 |
| April 1988 | 34.05 | 287 660 | 261 469 |
| May 1988 | – | – | 10 994 |
| June 1988 | – | – | 427 |
| Total | 95.30 | 434 564 | 415 773 |

## 3.4 Discussion

Recharge to the parts of the unconfined aquifers under consideration can originate from the following three sources:

1. Water flowing in the wadi bed just downstream of the dam (in the channel where the observation wells are located);

**Table 3.7**  Highest levels of water tabel (m) with respect to spillway crest and dates occurrence, Malham

| Season | Well | | |
|---|---|---|---|
| | M1 | S32 | S33 |
| 1985/1986 | 4.63 | 6.36 | 6.85 |
| | (28–4–86) | (28–4–86) | (1–5–86) |
| 1986/1987 | 12.39 | 11.81 | 11.53 |
| | (19–4–87) | (19–4–87) | (20–4–87) |
| 1987/1988 | 13.34 | 13.57 | 13.21 |
| | (12–5–88) | (20–5–88) | (19–5–88) |

**Table 3.8**  Highest levels of water table (m) with respect to spillway crest and dates of occurrence, Al-Amalih

| Season | Well | | |
|---|---|---|---|
| | S1 | S2 | S3 |
| 1985/1986 | 7.71 | 10.97 | 15.59 |
| | (17–4–86) | (19–4–86) | (21–4–86) |
| 1986/1987 | 8.50 | 11.65 | 16.18 |
| | (6–4–87) | (9–4–87) | (11–4–87) |
| 1987/1988 | 11.94 | 14.50 | 18.29 |
| | (27–4–88) | (29–4–88) | (3–5–88) |

2. Increase in the groundwater flow caused by water infiltrating into the wadi channel upstream (in the catchment);
3. Recharged volume ($R_v$) which infiltrates from the reservoir bed and joins the aquifer flow just upstream of the observation wells.

The first source is insignificant due to the fact that the length of the channel just downstream of the dams is short. The second can be very important, especially at Malham due to the large catchment under consideration. However, to show its separate effect on the water level in the downstream wells will require a comprehensive treatment of the whole aquifer, which is beyond the objective of this chapter. The third source ($R_v$) can be assumed to have a direct relationship to both the precipitation on the catchment and the water levels in the observation wells, and it is therefore a link between the runoff from the catchment (caused by precipitation) and recharge in the aquifer (indicated by a rise in water levels in the wells).

It is very clear from the results presented in Tables 3.1–3.6 that the greater the annual precipitation, the higher the value of $R_v$. The first season was the wettest among the three at both sites and large volumes of water were caught by the dams. Although the Malham catchment is more than thirteen times larger than the Al-Amalih catchment, the amount of inflow to the reservoir at Malham was less than twice that at Al-Amalih. It has been shown in another study (Al-Haqqan, 1989) that the runoff coefficient for Malham is much smaller than for Al-Amalih.

The recharge in the aquifers downstream of both dams is indicated by the rise of water levels, as shown in Figures 3.3 and 3.4. The highest levels reached by the water table at both locations are given in Tables 3.7 and 3.8. It can be seen that, at Al-Amalih, the wetter the year, the greater the water levels in the observation wells. For example, the annual precipitations (mm) for the three seasons were 175.45, 100.55, and 95.30, while the corresponding peaks of the water levels in well S1 (m) were 7.706, 8.498, and 11.936, respectively. Similar trends are also observed for the other wells at Al-Amalih. At the Malham site, the comparison between the peaks in the observation wells (Table 3.7) between the first and the second seasons and the annual precipitation follow the same trend as before. However, due to the small amounts of $R_v$ during the second and third seasons (compared to the first), the peaks in the latter were slightly lower than those in the former.

For the three seasons, it can be seen from Tables 3.1–3.6 that most of the annual precipitation occurs during the period from January to April, with a greater concentration in March and April. Most of the significant floods to the two dams (inflow) and recharge to the aquifer ($R_v$) occur during March and April. The dates of occurrence of the water table peaks in the observation wells (Tables 3.7 and 3.8) show that they were all in April or May.

## 3.5 Conclusions

After this study of the data available for the two sites under consideration, the following general comments can be made:

1. The annual precipitation will greatly influence the amounts of inflow reaching the dams and hence the amount of water available for recharge ($R_v$).
2. The most significant floods at the two sites studied occurred during March and April and the highest water table levels were in April or May.
3. For the region, most of the floods took place in March and April. This must be considered when devising management policies for recharge dams.

# References

Al-Dalooj, A. *et al.* (1983) 'Appraisal of recharge dams in the kingdom of Saudi Arabia', *Symposium on Water Resources in Saudi Arabia*, College of Engineering, King Saud University, Riyadh.

Al-Haqqan, A. A.(1989) *Analysis of Rainfall Runoff Data from Two Catchments in Central Saudi Arabia*, Senior Project Report, Civil Engineering Department, King Saud Univesity, January.

Al-Muttair, F. F. *et al.* (1986) 'Recharge characteristics from dammed wadis in central Saudi Arabia', in *Proceedings of International Conference on Water Resources, Needs and Planning in Drought Prone Areas*, Khartoum, Sudan, December.

Al-Turbak, A. S. and Al-Muttair, F. F. (1989) 'Evaluation of dams as a recharge method', *International Journal of Water Resources Development*, **5**, No. 2, June, 119.

Rizaiza, A., Allam, S. and Allam, M. N. (1989) 'Water requirements versus water availability in Saudi Arabia', *ASCE, Journal of Water Resources Planning and Management*, **115**, No. 1, January, 64.

*Water Atlas of Saudi Arabia* (1984) Ministry of Agriculture and Water, Riyadh.

# 4 CLIMATIC CHANGES AND TECTONIC ACTIVITIES AND THE DEVELOPMENT OF THE NILE DRAINAGE BASIN IN EGYPT

Abdu A. Shata
*Desert Institute, Cairo, Egypt*

## 4.1 Introduction

A wealth of information is available in the literature on tectonics and climatic changes and its bearing upon the evolution of the Nile drainage basin in Egypt (Figure 4.1) as well as in adjacent areas. This is found mainly in the work of geologists, geographers, archaeologists and hydrometeorologists (Amer and Huzaygin, 1952; Butzer, 1959; Said, 1962, 1981; Shata, 1962; Martin *et al.*, 1981; Klitsch *et al.*, 1984). This chapter is based on data from the published work as well as on field observations in the Nile Valley in Upper Egypt, in the Nile Delta in Lower Egypt and in the Eastern Desert of Egypt. *The Atlas of Egypt* (1928) was used in drafting the detailed outline of the catchment areas of the main wadis, which are morphologically developed on the eastern side.

## 4.2 Physical setting

The Nile drainage basin in Egypt falls in the arid zone belt of north-east Africa, where the rainfall decreases rapidly from north to south (an average of only about 100 mm characterizes the northern and eastern areas). On the eastern side of the basin, where there are the peaks of the Red Sea mountains, higher rainfall rates are recorded. Such peaks rise in places to more than 1500 m + MSL and are principally underlain by Basement rocks. In addition to such peaks, elevated structural plateaux characterize the landscape and constitute a part of the watershed area. The plateau areas are essentially underlain by carbonate rocks belonging to the Lower Tertiary. The main drainage lines dissecting the northern plateau areas are directed to the southern part of the Delta. On the western side of

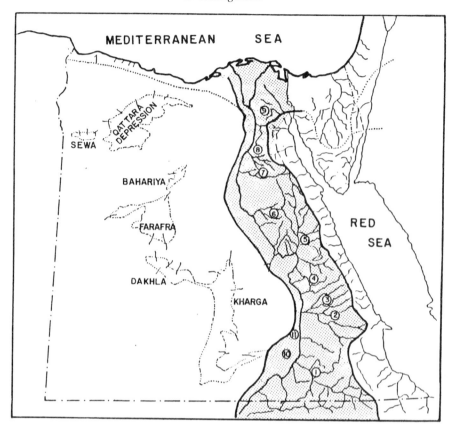

**Figure 4.1**  *Nile drainage basin*

the Nile drainage basin the landscape is much less complicated compared to the eastern side, and the ground elevation seldom exceeds 500 m + MSL. The surface is essentially underlain by carbonate rocks belonging to the Lower Tertiary in addition to the Upper Cretaceous. El Faiyum and Wadi El Rayan are two natural excavations (below sea level) or morphotectonical depressions in the Tertiary carbonate rocks. Although these two depressions have their own internal drainage systems, El Faiyum is connected to the Nile drainage basin at El Hawra Cut, which is traversed by Bahr Youssef (El Faiyum is connected to Wadi El Rayan by an artificial aqueduct). Wadi El Natrun is another natural excavation (below sea level), and is located outside the Delta basin to the west.

When considering the physical setting of the Nile drainage basin, it is of particular interest to give details of a number of selected catchment areas connected to the present Nile Valley. In order of magnitude, the catchment area of Wadi El Allaqui (Figure 4.2–A) is the largest known in the eastern side and is of the order of 20 000 km². From its uptake area in the Red Sea mountains beyond the border in Sudan to its mouth in the High Dam Lake to the south of Aswan the main trunk of Wadi El Allaqui has a length

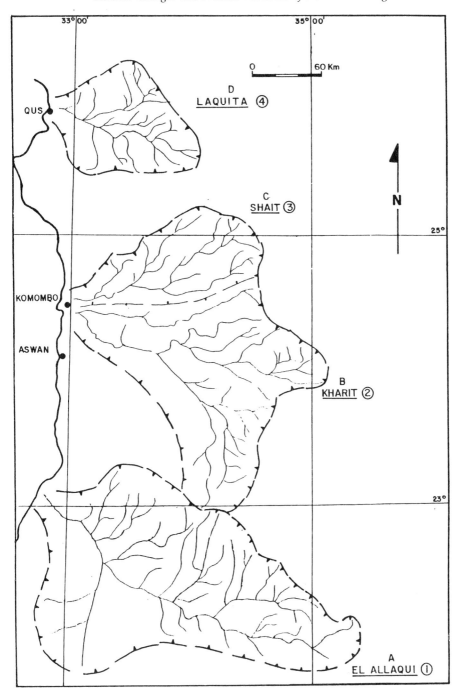

**Figure 4.2** *Southern main catchment areas*

of about 300 km and the rate of slope is about 0.6 m/km in the upstream part and about 0.07 m/km in the downstream one. The catchment area of Wadi Qena (Figure 4.2–A) is about 10 000 km². It is oriented north to south and the degree of slope is of the order of 0.6 m/km in the upstream part and about 0.3 m/km in the downstream one (about six times greater than Wadi El Allaqui). The catchment area of Wadi Tarfa (Figure 4.3) located further to the north and oriented east to west is about 4000 km². In the upstream part the rate of slope is about 0.6 m/km and is reduced to 0.2 m/km in the downstream one. Moving further north to the catchment areas of the wadis debouching in the Nile Valley in the area of Greater Cairo and in the southern part of the Delta, we have a complex system oriented north-west to south-east and occupying an area of about 4000 km². The catchment area of Wadi Gafra is nearly 50% the total and ends in the Nile Delta at Belbeis. In the upstream part the rate of slope is close to 0.6 m/km and is about 0.4 m/km in the downstream one.

On the western side of the Nile drainage basin the catchment areas of the wadis, best represented by Wadi Kalabsha and Wadi Kurkur, located to the south of Aswan (Figure 4.4), are smaller. Wadi Kalabsha has a catchment area of about 1000 km² and that of Wadi Kurkur is only about 400 km².

As mentioned above, the Nile drainage basin falls in the extremely arid belt of north-east Africa. This basin is, however, affected by occasional rainstorms (once every five or seven years), where the amount of runoff water exceeds 1000 million m³. Before the construction of the High Dam in the 1960s the author recorded rounded boulders and cobble gravels brought by Wadi El Allaqui and accumulated on the opposite side of the Nile. He also witnessed the colour of the water in the Nile at Cairo, changing during the flood season from light brown to milk-white, when the catchment areas on the eastern side which are incised in carbonate rocks became active.

## 4.3 A brief history of basin development

In the history of the development of the Nile drainage basin, occupying about one quarter of the area of Egypt, the following events are recorded and described in the literature:

1. *Later Miocene* (5.4 million BP); with regional epeirogenic land rising and a conspicuous lowering of the level of the Mediterranean. At that time the climate was humid and the early stages in the excavation of the Nile gorge as well as its tributaries (now desert wadis) occurred (Eonile).
2. *Early Pliocene* (3.32 million BP); with a gradual regional rise in the level of the Mediterranean and the formation of a marine gulf in the narrow gorge of the Eonile up to Aswan. Similar climatic conditions prevailed and the tributaries continued to flow into the Pliocene Gulf.

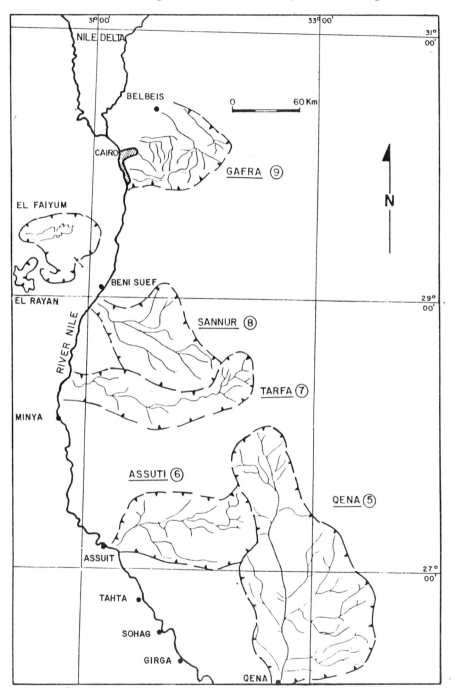

**Figure 4.3** *Northern catchment areas*

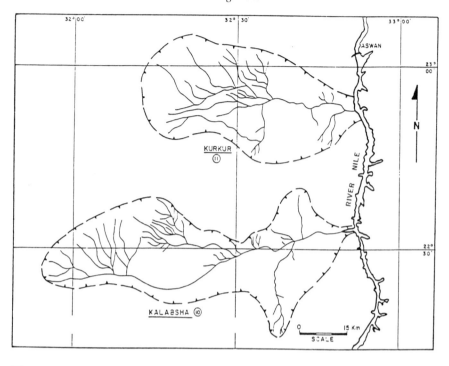

**Figure 4.4**   *Main catchment areas on the western side of the Nile*

3. *Later Pliocene* (1.86 million BP); another phase of land elevation, with the subsequent retreat of marine water and the development of a new river system (Paleonile). The climatic conditions were moist and the land surface was vegetated.
4. *Later Pliocene–Early Pleistocene* (1.71 million BP); a continued land elevation accompanied by seismic activities, with extremely arid climatic conditions interrupted by a short pluvial phase, which was characterized by winter rainfall and runoff water. The existing river system (Protonile) and its pre-existing tributaries occupied most of the present Nile drainage basin.
5. *Middle Pleistocene* (690 000 BP); the landscape in the Nile drainage basin remained unchanged and the river system became very active with a permanent supply of water brought from outside Egypt (Prenile). The climatic conditions were dominated by an arid phase in Egypt.
6. *Late Pleistocene* (120 000 BP); a landscape similar to the above, but with a great pluvial interval and heavy rain over Egypt. The main river system (intermittent) depended for its supply of water on the lateral derivatives coming mainly from the east (Neonile).
7. *Holocene* (25 000 BP); a landscape affected by the oscillation of Mediterranean sea level and dominant aridity. The present river system depends on its supply of water from outside Egypt (present Nile).

## 4.4 Climatic fluctuation in Holocene times

In view of the research work carried out in the Nile drainage basin as well as in the adjacent areas in north-east Africa and in south Asia, and which deals with aspects of isotope geology, archaeology, stratigraphy, biology, hydrology, etc., there is evidence which points to climatic fluctuations, with wet intervals, in Holocene times (beginning about 25 000 BP). However, the wetter intervals recorded in Holocene times are not taken as a repetition of the Late Pleistocene. The following is a brief account of events:

1. A severe desertification phase took place from about 11 500 or 10 000 BP. The upper limit of this phase coincides with the Late Weichselian/ Falandrian boundary of West Europe (8000 BP).
2. A sub-pluvial phase occurred towards the end of early Holocene times (about 7000–5000 BP). This phase is referred to as the 'Neolithic sub-pluvial'. During this interval, the Blue Nile flood volume was considerable and accumulation of the silt deposits was manifested in the valley (as a response to the gradual rise in the level of the Mediterranean in post-glacial times, silt accumulation began much earlier to the south). This sub-pluvial phase has left its impressions on the eastern side of the Nile drainage basin, where the wadi beds were degraded at least 15 m and also in recorded animal remains and drawings (Figures 4.5 and 4.6).
3. A phase with declining and fluctuating rainfall marked the termination of the sub-pluvial phase (from about 4350 BP, which is equivalent to the Sixth Dynasty). At about 4000 BP (Old Kingdom) the significant effects of invading sands and receding floods became dominant in the Nile drainage basin south of Cairo (Figure 4.7).

   The dunes are reported to have 'invaded the Nile Valley in Western middle Egypt, probably covering a 175 km stretch of former alluvium to a depth of 0.5 km by many metres of sand'. This aeolian phase may correspond to the second desertification phase reported in south Asia at about 3800–3500 BP.
4. Another wet phase, characterized by the recovery of vegetation and extensive flooding took place in the New Kingdom and Late Dynastic times (about 2500 BP). According to the testimony of Herodotus (450 BC) and of Roman authors in the first century, strong floods in the Nile 'achieved 9.54 m as a rule, which is 50 to 100 cm higher than was the case in Islamic times'.
5. The last phase, characterized by relatively stable climatic conditions, has been common in the past 2000 years. However, there are records which designate several episodes with climates, which 'are significantly different from present'. A recent episode occurred in about the sixteenth to eighteenth centuries, and was probably synchronous with a period of glacial advance and frequent severe winters in Europe. This is generally referred to as the Little Ice Age. During at least the early part of this

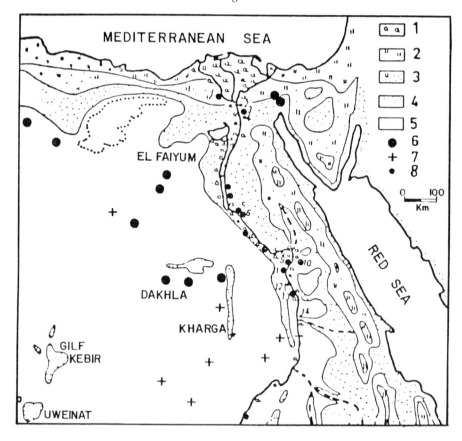

**Figure 4.5**  *Early hunters and cattle-nomads in the Sahara in Neolithic times (after Butzer). 1 Extent of Hamitic livestock-raising culture (4000 BC); 2 extent of eastern hunters (4000 BC); 3 domain of central group of hunters; 4 rock-drawings of early hunters; 5 rock-drawings of early hunters and cattle-nomads; 6 rock-drawings of cattle-nomads; 7 hypothetical routes followed by cattle-nomads*

phase, extensive empires flourished in places located south of the Sahara in North Africa, such as the Hausa States in North Nigeria and the Songhay Empire in North Mali. By the nineteenth century, a climate similar to the present one had been established, although with a brief return to more humid conditions. Information from North Africa (Senegal, Niger, Chad and the Nile) indicates that in the twentieth century three major drought intervals occurred: in 1913, 1940 and 1972 (Figure 4.8). In the Nile drainage basin the desertification processes associated with the last drought (1972) are particularly felt in the encroachment of shifting sands on the cultivated land areas between Minia and Assuit in Upper Egypt.

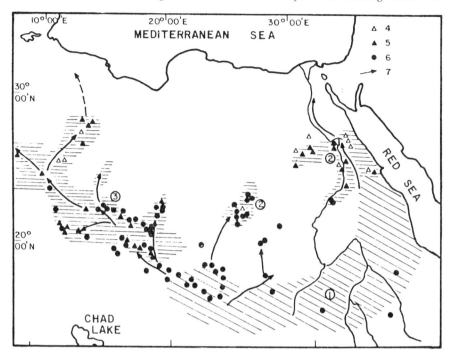

**Figure 4.6**  *Climatic-ecological vegetation zones in Egypt, 5000–2000 BC (after Butzer). 1 Galeria woodland; 2 good pasture with over 150 mm precipitation; 3 moderate pasture with 100–150 mm precipitation; 4 pasture after rain, 50–100 mm precipitation; 5 negligible pasture, less than 50 mm precipitation; 6 large oases; 7 larger watering places; 8 prehistoric sites*

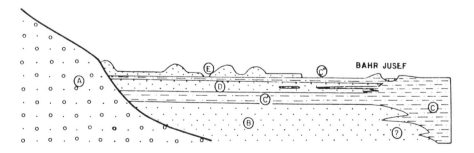

**Figure 4.7**  *Section of the marginal valley dunes (after Butzer). A Pleistocene gravels; B older dunes (350–500 BC); C nilotic silt; C' nilotic silt, Graeco-Roman period; C" nilotic silt, Islamic period; D lower younger dunes, Byzantine to Islamic period; E upper younger dunes, sub-recent*

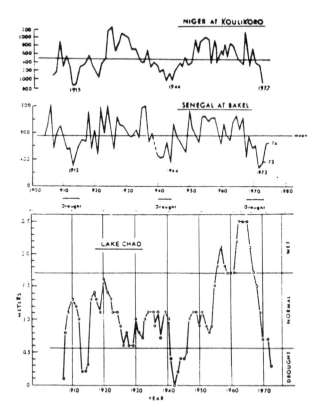

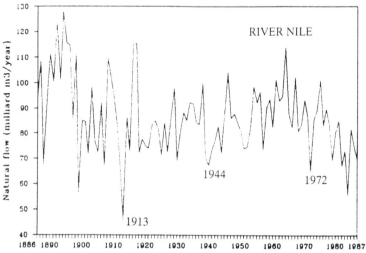

**Figure 4.8**   *Comparisons of the annual fluctuations in water discharge of the rivers Niger, Senegal and Nile and maximum annual levels of Lake Chad. The major droughts in 1913, 1944 and 1972 are clearly seen on all curves*

# References

Amer, M. and Huzayyin, S. A. (1952) 'Some physiographic problems related to the Predynastic site of Maadi', *Proc. Panafrican Congr. Prehist.*, Oxford University Press, Nairobi, 1947.

Butzer, K. W. (1960) 'Environment and human ecology in Egypt during Predynastic and Early Dynastic times', *Bull. de la Soci. de Geogr. d'Egypt.*

Jaiswal, P. L. *et al.* (1977) *Desertification and its Control*, Indian Council of Agricultural Research, New Delhi.

Klitsch, E. *et al.* (1984) *Research in Egypt and Sudan*, Verlag von Dietrich Peimer, Berlin.

Martin, A. J. W. *et al.* (1980) *The Sahara and the Nile Quaternary Environments and Prehistoric Occupation in Northern Africa*, Balkema, Rotterdam.

Rapp, A. *et al.* (1976) 'Can desert encroachment be stopped? A study with emphasis on Africa'. *Ecological Bull.* **24**.

Said, R. (1962) *The Geology of Egypt*, Elsevier, Amsterdam.

Said, R. (1981) *The Geological Evolution of the River Nile*, Springer-Verlag, New York.

Shata, Abdu, A. (1962) 'Remarks on the geomorphology, pedology and groundwater potentialities of the southern entrance to the New Valley', *Bull. de la Soci. de Geogr. d'Egypte*, **35**.

# 5 EGYPT'S PROGRAMMES AND POLICY OPTIONS FOR LOW NILE FLOWS

Mahmoud A. Abu-Zeid and Safwat Abdel-Dayem
*Water Research Center, Cairo, Egypt*

## 5.1 Introduction

For Egypt the Nile is more than just a river. It is a whole socio-economic system which has turned one of the most arid regions of the world into a prosperous valley with rich agricultural production. The rainfall in Egypt is very scarce and scattered. It is limited to the coastal areas during the winter season with an intensity not exceeding 200 mm/yr. The intensity decreases to 50 mm/yr near Cairo and to almost zero further to the south. Even the groundwater aquifers under the Delta and the Nile Valley are replenished by deep percolation of the Nile water. Therefore life in Egypt is completely sustained and supported by this water. Irrigation, industrial, municipal, navigation and hydropower generation needs are all covered by the Nile.

An important characteristic and typical feature of the Nile is the fluctuation of its flow. The climatic conditions over its catchment area determine how much water will be available every year. The quantity of rainfall and the evaporation losses vary over the year and from one year to the next. Conditions of high floods and very low flows may occur as individual incidents or as repeated events in cycles. Forecasting of the flows and their variation is essential for development and proper planning. Moreover, it is necessary for avoiding the damaging effects of severe droughts and disastrous floods.

## 5.2 The Nile flow and present forecasting procedure

The Nile water originates from the Ethiopian plateau in east Africa and the equatorial lakes plateau in central Africa. The annual flow of the river depends largely on the intensity of rainfall over the two plateaux. Water losses through evaporation, seepage and spillage in the equatorial lakes plateau are too high, and therefore most of the water arriving at Aswan originates from the Ethiopian plateau (Abu-Zeid and Abdel-Dayem, 1990). About 85% of the annual natural flow of the Nile arriving at Aswan comes from the Ethiopian plateau while only 15% of the flow is from the equatorial plateau.

The natural annual flow of the Nile varies from one year to the next. The long-term average flow at Aswan was determined as 84 milliard m³/yr. However, flow as high as 127 milliard m³/yr sometimes occurred, while a very low flow of 46 milliard m³/yr took place during the nineteenth century. The annual flow hydrograph at Aswan (Figure 5.1) shows the fluctuation of the Nile during the period 1886–1987. A decreasing trend in the moving average flow can be noted over this period. The high flows characterized at the end of the last century brought the annual average of the period 1876–1900 up to 102.5 milliard m³/yr. Recently the annual flows are generally low to the extent that the average flow of the period 1972–1987 is only 78.4 milliard m³/yr. The frequency distribution of the natural annual flow during the period 1900–1987 is shown in Figure 5.2. The long-term average flow during this period is 83.89 milliard m³/yr.

The seasonal and annual fluctuation of the Nile's natural flow may cause great problems for a country such as Egypt, which depends completely on the river. Early information about the river's flow was always necessary for short-term planning to determine maximum attainable crop intensities, flood-control measures or policies for facing shortages and drought conditions. In the absence of long-term storage before the construction of the High Aswan Dam (HAD), the situation was more sensitive and vulnerable to extreme variations. Many disasters occurred as a result of extremely high or low floods.

Forecasting of flood conditions was traditionally based on streamflow data. A network of gauging stations was established along the Nile Basin to

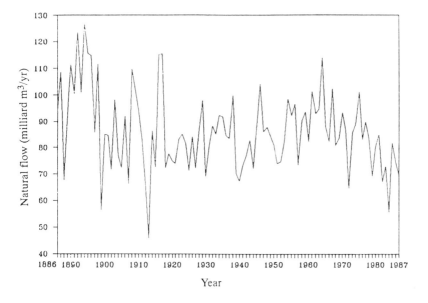

**Figure 5.1** *Hydrograph of the Nile's natural flow, 1886–1987*

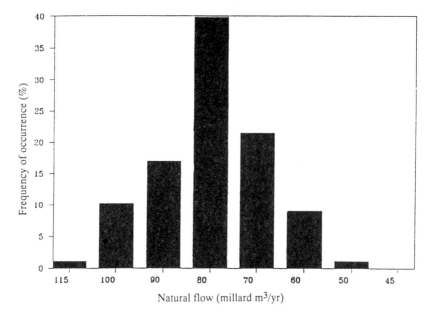

**Figure 5.2**  *Frequency distribution of the Nile's natural annual flow during the period 1900–1987*

measure water levels and discharges. The records at one station were used to determine the travel time and the magnitude of flow at the other downstream stations (Fahmy, 1960). The success of this routing technique in predicting the flow conditions at any point along the stream depends on the accuracy in calculating the travel time and the water losses within the reach considered. In spite of the short lead time given by this flood-forecasting method, it provided great help, especially in situations where safety measures were required to counter high floods.

In order to face the uncertainties associating the Nile's natural flow, many engineering storage, control and regulation projects have been implemented during the past 100 years. Works were constructed on the Nile and its tributaries for water storage and conservation. The first storage work built in comparatively recent times in Egypt is the old dam at Aswan (1898–1902), to store just 1 milliard m$^3$ of the floodwater and to use it together with the natural supply of the river in the following low-flow season. Its capacity was raised to 5.0 milliard m$^3$ in 1934 by increasing its height. The Sennar Dam was completed in 1925 on the Blue Nile for the benefit of the Sudan. The Jabal el Awlia Dam was constructed in 1937 on the White Nile to store water for Egypt. Its storage, which was initially 3.5 milliard m$^3$, is decreasing due to continuous silting up. The storage of the Khashm el Girba Dam, which was constructed in 1960–1964 with an initial capacity of 1.3 milliard m$^3$ to regulate some of the Atbra water, is also decreasing. In 1961–1966 the Roseires Dam was built on the Blue Nile to retain a maximum storage of 7.6 milliard m$^3$ of water for the Sudan.

Long-term storage work in the Nile Basin started with the construction of the Owen Falls Dam. Its objective was to increase the Lake Victoria storage by 200 milliard $m^3$. Due to the high losses in the Sudd region and the Bahr el Jebel Basin, any improvement in the lake outflow to the Sudan or Egypt becomes questionable unless the river channels are equally improved (Shahin, 1985). The HAD was constructed on the main river south of Aswan and became operational in 1964. It is designed to allow a dead storage of 30 milliard $m^3$ and a live storage of 90 milliard $m^3$ in its reservoir and an additional emergency capacity of 41 milliard $m^3$ before the Toshka spillway starts flowing. Shares of Egypt and the Sudan in its water storage were fixed by the 1959 Nile Water Treaty at 55.5 and 18.5 milliard $m^3$/yr, respectively.

A series of control works was constructed on the main Nile between Aswan and the Mediterranean. The Mohammed Ali Barrages were the first control work completed in 1861 at the apex of the Delta. These were followed by other barrages at Esna, Naga-Hammadi, Assiut, Zifta and Edfina. New barrages were built in 1939 to replace the Mohammed Ali Barrages which could not withstand further rebuilding. The barrages are open-type dams, where water flows through their vents. The flow through the vents as well as the water levels upstream and downstream of the structure are regulated by means of vertical steel gates.

## 5.3 The latest drought and the HAD storage

The HAD is an engineering feat which has proved to be the saviour of the nation against one of the most severe drought cycles known in modern history. Since the completion of the HAD, its reservoir levels were gradually rising to reach almost its maximum live storage in November 1978. The next nine years (1979–1988) witnessed a chain of low floods due to the drought conditions prevailing in the Horn of Africa. Water arriving at Aswan annually was always less than the necessary releases for Egypt's needs. In 1984–1985 it was only 35.6 milliard $m^3$/yr. As a result, the water levels in the reservoir gradually dropped to alarming levels (Abu-Zeid and Abdel-Dayem, 1990). In July 1988 a minimum level of +150.60 m was recorded, which corresponded to a total storage of 38.4 milliard $m^3$.

Over the nine years of low floods the HAD reservoir was sufficient to avoid any negative impact of the water shortage on Egypt's economy. However, due to difficulties in making long-term forecasts, the government prepared emergency plans to counter any further drought conditions during the following years. These included reduction in the annual releases through more efficient regulation, extension of the irrigation system's winter-closure period to 4 weeks instead of 3 weeks, reduction of the area under rice to 900 000 acres and improving the Nile's navigable channel to reduce the water releases required to maintain the minimum required draught during the period of lowest irrigation demands.

Heavy rains unexpectedly occurred during the early summer of 1988 over the Ethiopian plateau, causing one of the highest floods. The total annual flow arriving at Aswan during the water year 1988–1989 was about 90.3 milliard m³. The reservoir refilled and reached a maximum elevation of +168.69 m in December 1988. The storage capacity of the reservoir corresponding to this elevation was 92 milliard m³. The flow of this year was thought to indicate an end of the drought period and a beginning of a cycle of high flows. Obviously, a forecasting technique with a sufficiently long lead time was lacking. The flood of the year 1989–1990 was disappointingly below average. The total flow arriving at Aswan during the period July–October 1989 was only 39.9 milliard m³ against 70.32 milliard m³ in the previous year.

The next section of this chapter will discuss a number of plans and studies made by the Ministry of Public Works and Water Resources (MPWWR) to predict and counter future Nile flows. The aim is to develop a means of obtaining reliable forecasting with a sufficient lead time for more efficient management of the Nile's water. Technical studies and institutional arrangements have also been considered to establish a powerful tool in the MPWWR to achieve this objective. However, an immediate aim was to use the mathematical models developed over the last few years to determine several options for facing all possibilities.

## 5.4 Prediction of future river flows

The critical situation imposed by the continuous low floods of 1979–1988 required an immediate evaluation of an uncertain future. The MPWWR issued a study in May 1988 to determine several options for operating the HAD reservoir under all possible Nile flows. One objective of the study was to ensure Egypt's annual water needs for agricultural, municipal, industrial and navigational uses. Another was to regain the safe storage volume lost over the drought years.

The 'Lake Nasser' simulation model was used to predict the reservoir levels and storage for different scenarios of river flows reaching HAD and a range of possible releases downstream of the Dam. The simulation was made for two assumptions of reservoir annual levels. The first was to have the same level of +150.00 m at both the beginning and end of the water year. The second was to start with an elevation of +150.00 m and to reach a level of +147.00 m at the end of the year. These operational levels were maintained for five different discharges of downstream releases, ranging between 55.5 and 50.5 milliard m³/yr, to determine the corresponding natural river flows. This range of releases covers the different possibilities for operating the Dam under all expectations of meeting water needs. The predicted natural flows for the two reservoir elevation requirements are given in Table 5.1, with annual variations of reservoir levels as shown in Figures 5.3 and 5.4, respectively.

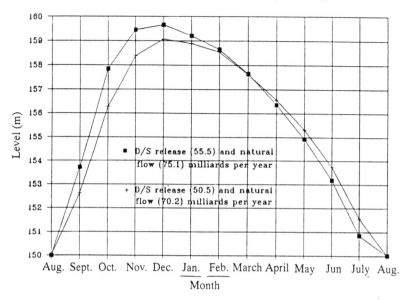

**Figure 5.3** *Predicted annual variations of levels upstream of the HAD when starting and ending elevations of +150 m are maintained*

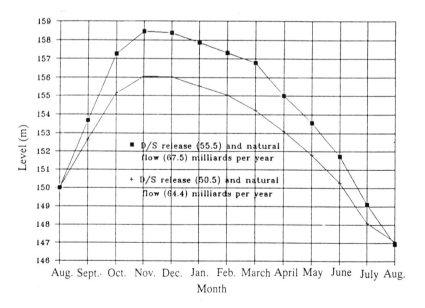

**Figure 5.4** *Predicted annual variations of levels upstream of the HAD when the level is +150 m at the beginning and +147 m at the end of the water year*

The basic requirements considered in simulating the different operation policies were as follows:

1. No reduction in municipal and industrial water requirements.
2. Maintaining a draught of 90 cm at the Nagah-Hammadi lock in December and January.
3. Ensuring water requirements for reclaiming about 150 000 acres annually that were not irrigated previously.
4. The only accepted change in the crop pattern to be the reduction of areas under rice which could be replaced by maize.

**Table 5.1**   Predicted natural flow to maintain certain reservoir level requirements

| Flow condition | D/S releases | Flow arriving at Aswan (milliard $m^3$/yr) | |
|---|---|---|---|
| | | Assumption I[a] | Assumption II[b] |
| High | 55.500 | 75.100 | 67.500 |
| Above average | 53.375 | 73.500 | 65.300 |
| Average | 52.500 | 72.400 | 65.900 |
| Below average | 51.500 | 71.400 | 64.800 |
| Low | 50.500 | 70.200 | 64.600 |

[a]Starting and ending annual reservoir levels are +150 m.
[b]Starting reservoir level is +150 m, ending at +147 m.

As it is extremely difficult to forecast the size of the Nile flood before the end of August every year, it was necessary to consider a number of possibilities based on the historical data of the floods. The records of the river's natural annual flow arriving at Aswan between 1900 and 1987 were used to identify five classes of river flows, and can be described as shown in Table 5.2.

The 'Lake Nasser' simulation model was used again in this case to predict the reservoir water levels throughout the year in each case. The assumptions made were to consider a reservoir level of +150.00 m at the

**Table 5.2**   Representative values of the Nile's annual flow during the period 1990–1987

| Flow condition | Water years | Annual flow (milliard $m^3$/yr) |
|---|---|---|
| High | 1916–1917 | 115.2 |
| Above average | 1938–1939 | 100.0 |
| Average | 1926–1927 | 84.1 |
| Below average | 1971–1973 | 64.6 |
| Low | 1984–1985 | 55.7 |

beginning of the year and annual releases varying between 55.5 and 50.5 milliard m³/yr. The levels' variation were predicted for each of the years in Table 5.2 and the following two years, according to their recorded inflows. It was assumed that any of these series may be repeated during the period 1988–1991. Thus the objective of this analysis was to determine the effect of different inflows and release policies on lake storage and its impact on agricultural and hydroelectric power generation needs.

The results of this study revealed that reservoir levels will be always higher than +150.00 m during the successive three years when high, and above-average and average inflow occur during the first year in the series. Reduction in levels below the +150.00 elevation may take place during some months only with below-average and low floods. The number of months per year during which the levels are expected to be lower than +150.00 and +147.00 m during the 3-year time series are given in Tables 5.3 and 5.4.

**Table 5.3** Number of months during which reservoir levels will drop to critical values[a] for below-average inflow conditions

| Annual releases (milliard m³/yr) | Levels below +150 m | | | Levels below +147 m | | |
|---|---|---|---|---|---|---|
| | Year 1 | Year 2 | Year 3 | Year 1 | Year 2 | Year 3 |
| 55.500 | 3 | – | – | 2 | – | – |
| 53.375 | 3 | – | – | 1 | – | – |
| 52.500 | 3 | – | – | 1 | – | – |
| 51.500 | 2 | – | – | 1 | – | – |
| 50.500 | 1 | – | – | – | – | – |

[a]Critical levels occur during the last months of the water year August/July.

**Table 5.4** Number of months during which reservoir levels will drop to critical values for low inflow conditions

| Annual releases (milliard m³/yr) | Levels below +150 m | | | Levels below +147 m | | |
|---|---|---|---|---|---|---|
| | Year 1 | Year 2 | Year 3 | Year 1 | Year 2 | Year 3 |
| 55.500 | 6 | 4 | 4 | 4 | 3 | 3 |
| 53.375 | 6 | 3 | 2 | 4 | – | – |
| 52.500 | 6 | 3 | 2 | 4 | 1 | – |
| 51.500 | 6 | 3 | 1 | 3 | 1 | – |
| 50.500 | 6 | 2 | – | 3 | – | – |

The monthly reservoir levels for each inflow and corresponding outflow were determined together with the live storage. It was found that all agricultural demands will be covered except for rice, when the annual releases become less than 53.375 milliard m³/yr. The area under rice should be decreased to 1.024, 0.888 and 0.751 million feddans when the downstream releases are 52.5, 51.5 and 50.5, respectively. This would ensure the water required for annual reclamation of 150 000 acres.

The needs for power generation could be estimated through the predicted reservoir levels and monthly releases from the reservoir. It was found that reduction in power generation would be necessary only when the reservoir level drops below +147.00 m. This would occur mostly for one month at the end of the first year, when the natural river flow is below average (Table 5.1). However, when the annual releases are decreased to 50.5 milliard m$^3$/yr, reduction in electric power generation would not be necessary. The production can be increased when this happens through decreasing the agricultural areas under rice to less than 700 000 acres annually.

## 5.5  The Hydrological Forecasting and Monitoring Center

For the improvement of flood prediction, the MPWWR is working towards establishing a centre for forecasting short-term and long-term variations of the river flow in real time. The centre would be predominantly in charge of determining the inflows to the HAD reservoir. To this end, a 'Planning Studies and Models' department was established within the Water Planning Group, to be responsible for monitoring, forecasting and simulation of the Nile. This is aimed at producing both deterministic forecasts and extended streamflow predictions (MPWWR, 1988). The deterministic forecasts will start at a specified time and initial state of the Nile Basin and will compute the hydrograph that would be produced by a given pattern of precipitation. The extended streamflow predictions will be based on replicating the deterministic forecasts a specific number of times, each using a random precipitation pattern. The set of hydrographs will be used to compute future probabilities during a specified period.

The techniques adopted will give a longer lead time than is available from the present forecasting procedures based solely on streamflow data. This increased lead time should be gained through simulating the upper basin's response to rainfall, accounting for the moisture balance of the soil–groundwater system and long-range climatic signals-based precipitation forecasts. In order to achieve these capabilities the system will include data acquisition, remote sensing, hydroclimatic analysis, diagnosis and prediction.

## 5.6  Conclusion

The High Aswan Dam (HAD) was constructed to ensure a stable water supply for both Egypt and the Sudan. However, the drought period which occurred from 1979 to 1988 proved that a relatively long series of low floods of the Nile should be expected. However, even with a full reservoir content, there will be a risk of not meeting the water requirements under normal operating conditions of the Dam when a long drought cycle takes place.

When the reservoir contents continued to diminish with the continuation of low floods in 1979–1988, Egypt implemented immediate plans to counter the critical situation. The actions which were considered for saving water included a decrease in annual releases, extension of the winter-closure period and a reduction in the intensity of the rice crop. Computer simulations of the inflow into the HAD reservoir have been used to predict the impact of different operational alternatives on water requirements. Studies of this type proved that power generation from the HAD may be affected by low and below-average floods. The study also showed that the intensity of the rice should be decreased when the annual releases from the dam are less than 53.375 milliard m$^3$/yr, to ensure the water required for land reclamation.

The floods of the following years were not according to expectations. While a high one occurred during 1988, a below-average flood took place in 1989. Therefore the need for reliable forecasting procedures became necessary to deal with fluctuations in the annual flow. The present forecasting procedure depends on monitoring the flows and levels at a number of stations along the Nile, from its sources to Aswan. This technique permits a lead time of not longer than one month. The Ministry of Public Works and Water Resources have established a centre for forecasting the variations of the river flow with a longer lead time. This increased lead time will be realized through using remote sensing for monitoring the hydroclimatic system of the upper basin. Accounting for changes in soil moisture and groundwater systems will improve the accuracy of the forecasts.

# Acknowledgement

The authors are grateful to Dr B. B. Attia for his assistance and cooperation, which greatly helped in the preparation of this chapter.

# References

Abu-Zeid, M. A. and Abdel-Dayem, S. (1990) 'The Nile, the Aswan High Dam and the 1979–1988 drought', *14th International Congress on Irrigation and Drainage (ICID)*, Vol. I-C, Q43, R26, Rio de Janeiro, Brazil.

Fahmy, H. K. (1960) *Forecasting the River Flows and Elevations*, Ministry of Public Works, General Inspectorate for Nile Control, Egypt (in Arabic).

MPWWR (1988a) 'Expectations and possibilities of Nile flow during the next three years (1989–1992) and how to ensure Egypt's water needs', Internal Report (in Arabic).

MPWWR (1988b) 'Monitoring, forecasting and simulation. Irrigation systems management, planning studies and models component', (Final Draft) Workplan.

Shahin, M. (1985) *Hydrology of the Nile Basin*, Elsevier Science, Amsterdam.

# 6 THE HYDROLOGICAL AND METEOROLOGICAL FACTORS CONTROLLING FLOOD FORECASTING

S. H. Sharaf El-Din, A. A. Khafagy and A. M. Mahar
*Water Research Center, Coastal Research Institute, Egypt*

## 6.1 Introduction

In flood forecasting, meteorological information could provide an early warning of floods. Much of the rain which falls in mountainous regions such as the Ethiopian highlands is directly due to the elevation of moist air over the mountains. The rainfall over Ethiopia originates from moisture brought by air currents from the south-west and caught by the Ethiopian scarp and highlands. The largest proportion of runoff is due to rainfall on steep slopes.

It seems, however, that the rainfall over Ethiopia resembles (in a less intensive way) the Indian south-west monsoon, shedding its precipitation along the more than 1000 m high scarp parallel to the coast. Also in Ethiopia a large part of the runoff originates on the mountain slopes facing west.

Ethiopia can be divided into three major zones according to its topography. The greater part is at 1800–2400 m elevation and comprises undulating plateaux. From the plateaux, mountains and peaks rise to above 4000 m. The Blue Nile and Atbara discharges are also governed by the Ethiopian monsoonal rainfalls.

Relating the rainfall to meteorological parameters such as humidity, pressure and wind distribution over the Ethiopian highlands, flood forecasting for the Nile could be carried out.

The object of this chapter is to predict the rainfall using meteorological parameters such as pressure and moisture content over the Ethiopian highlands. A linear regression equation was derived to satisfy a prediction for both the rainfall and the Nile flood discharge. The data used in these analyses are those of the rainfall on the Ethiopia highlands and the corresponding Nile flood discharge for the period 1952–1970.

## 6.2 The distribution of meteorological parameters over the Ethiopian highlands

The meteorological parameters over the Ethiopian highlands such as pressure, wind, temperature, humidity and precipitation can be described as follows:

1. *Pressure.* The seasonal reversal of the pressure pattern over the Ethiopian highlands is shown in Figures 6.1–6.4. Winter distribution is dominated by a trough of low pressure from central Africa. The situation differs in summer when the area is covered by a belt of relatively high pressure. The summer pattern prevails during June to August. The autumn transition occurs in September and October, when the area is affected by a trough of low pressure. The spring change is gradual during February and March, and then more rapid in April and May. Variations from average occur in all months and cause significant local weather changes.
2. *Winds.* Wind observations over the region show great regularity throughout the year. The topography and the large diurnal range of temperature have a major influence on wind direction and speed. The mean wind speeds are 7.2 knots (winter), 6.7 knots (spring), 6.6 knots (summer) and 7.5 knots (autumn).

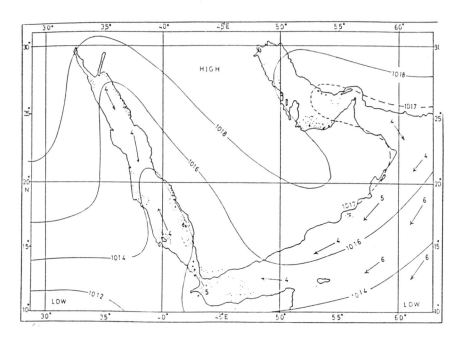

**Figure 6.1**   *Mean barometric pressure (mb) and dominant winds (mean force) –* *January*

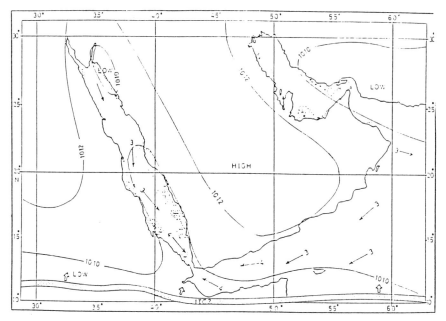

**Figure 6.2** *Mean barometric pressure (mb) and dominant winds (mean force) –*
*March*

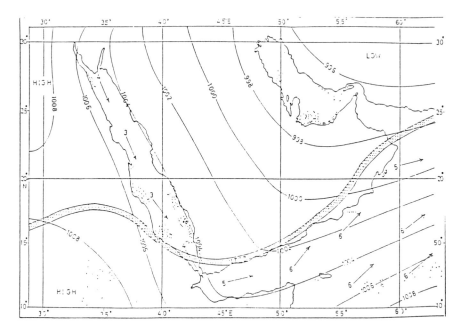

**Figure 6.3** *Mean barometric pressure (mb) and dominant winds (mean force) –*
*July*

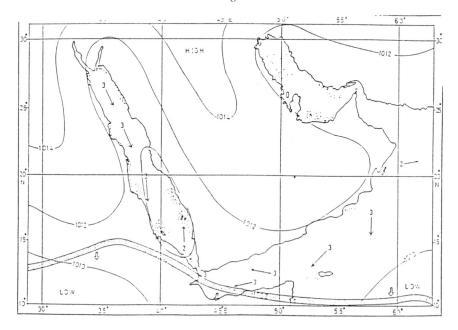

**Figure 6.4**   *Mean barometric pressure (mb) and dominant winds (mean force) –*
*October*

3. *Air temperature.* The average air temperature in February is 18°C, rising
   to a maximum of 43°C in August. Sudden temperature changes are
   encountered in the disturbed convergence zone.
4. *Humidity.* Average values are about 68% or less. Maximum values occur
   around dawn, with readings of more than 80%. The prevailing high
   temperature over the whole region tends to obscure the actual vapour
   content. The minimum humidity occurs in summer (56%), while the
   maximum is encountered in December and January (76%).
5. *Precipitation.* Rainfall occurs from May to September and is mainly due
   to the proximity of the mountainous terrain. Most occurs at the
   convergence zone. Some heavy falls due to thundery rainstorms are
   recorded to be 90 mm in one day. Snow sometimes covers the peaks of
   the higher mountains in winter while thunderstorms occur during the
   period May to December.

## 6.3 Rainfall on the Ethiopian highlands for the period 1952–1970

The annual and the average monthly rainfall over the Ethiopian highlands
are shown in Figures 6.5 and 6.6. From these figures it is clear that there is
a significant increase in rainfall during July and August. September and

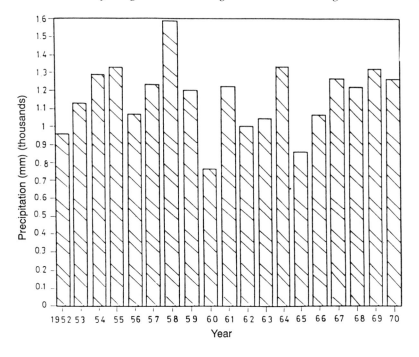

**Figure 6.5**  *Histogram of rainfall in Ethiopia*

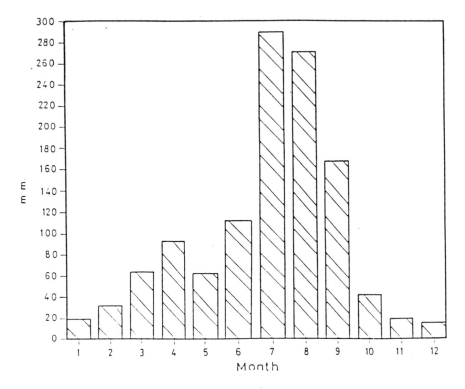

**Figure 6.6**  *Histogram of rainfall in Ethiopia from 1962 to 1970*

June have higher values than the remaining months. The average annual rainfall over the Ethiopian highlands amounts to 1230 mm. Figure 6.5 shows that 1958 had the maximum amount of rainfall, while 1960 and 1965 had the minimum. The rainy season over the Ethiopian highlands begins in May and ends in September (Figure 6.5).

## 6.4 Nile discharge during the period 1952–1970

Figures 6.7 and 6.8 indicate the annual and average monthly Nile discharge, respectively. The mean annual Nile discharge is $84 \times 10$ m. From Figure 6.7 it can be seen that the maximum discharge occurs in 1964. Figure 6.8 shows that the Nile discharge at Aswan begins to increase in June and reaches a maximum in September. Study of the mean monthly Nile discharge indicates that the rains take about one month to effect a flood at Aswan.

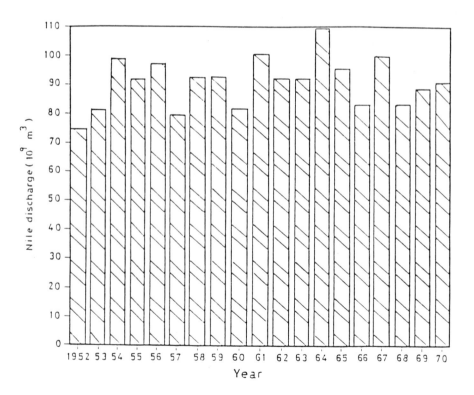

**Figure 6.7**   *Histogram of Nile discharge from 1952 to 1970*

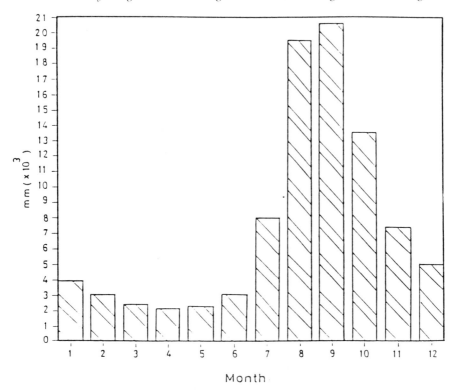

**Figure 6.8** *Histogram of Nile discharge over a twelve-month period*

## 6.5 Models for flood forecasting using meteorological parameters

Meteorology provides a method of obtaining the earliest warning of floods. By using meteorological parameters such as atmospheric pressure and humidity, the maximum amount of rainfall can be estimated. For forecasting rainfall, the pressure and the humidity are averaged over a 5-year period (1970–1974).

The maximum precipitable water vapour ($W$) of a column of air is given by

$$W = \frac{1}{g} \int_{1}^{2} m \, dP \tag{6.1}$$

where

$$m = \frac{0.622e}{P - e}$$

where $P$ is the atmospheric pressure (mb), $m$ is the saturated mixing ratio (g/kg) and $e$ is the saturated vapour pressure (mb). The graphical solution of equation (6.1) (George and Frank, 1957) yields

$$W = \frac{\overline{m}(P_1 - P_2)}{g}$$

where $\overline{m}$ is the mean saturated mixing ratio of the column of the air, $P_1$ is the surface atmospheric pressure, $P_2$ is the pressure at the surface where the water vapour is minimum and $g$ is the earth's acceleration. The pressure is derived from synoptic weather charts while the water vapour content is obtained from climatic tables.

   The predicted Nile flood discharge can be calculated from the surface wind speed, the water vapour content and the catchment area together with the height of the air column from the surface to the point of minimum water vapour content. The relationship between these parameters is given by (Brunt, 1952)

$$R = 2\pi \; h.u.r.\overline{m} \; \text{mm/s}$$

where

> $R$ is the rate of precipitation over the catchment area,
> $h$ is the height of the air column (m),
> $u$ is the surface wind speed (m/s),
> $r$ is the radius of catchment area (m), and
> $m$ is the mean mixing ratio in the air column (g/kg).

## 6.6 Comparison between observed and predicted rainfall and Nile discharge

With the above equations, the predicted rainfall over one year was calculated using the observed rainfall data over 19 years, while the meteorological data were taken over the 5 years from 1970 to 1974. For rainfall a regression line was calculated for all the observed and predicted values in three cases:

1. All the observed values against all the predicted ones;
2. All the observed values of less than 100 against the corresponding predicted ones;
3. All the observed values of more than 100 against the corresponding predicted ones.

The equations for the regression lines are as follows:

1. $Y = 24.37 + 0.486X$
2. $Y = 9.184 + 0.795X$
3. $Y = 68.132 + 0.298X$

where $X$ is the observed value and $Y$ the predicted one.

Using the above three models, the result showed that regression line (2) fits more accurately with all observations rather than the other two. This indicates that the model can be applied with a high degree of accuracy for the period from February to October but it fails for the months of low rainfall. The data and results are shown in Table 6.1.

In general, it can be concluded that the model can be applied for reasonably accurate prediction whose error is within 15% which is considered to be satisfactory for such a prediction. Also, the total predicted rainfall over the year is 1043.4 mm, which is 130.6 mm less than the observed one with a percentage error of 11.1.

In the case of fresh water discharge only one regression line was calculated, which gives the result shown in Table 6.2. The equation for this is

$$Y = 2.014 + 0.869X$$

where $X$ is the observed value and $Y$ the predicted one.

The results show that the model can be applied with a high percentage of accuracy in the months of high discharges. Other months have a smaller percentage. This can be explained on the basis that the prediction is deduced from meteorological parameters which are suitable for producing rain without regard to others such as stability, temperature, dewpoint and wind direction, which are very effective in determining the rate of precipitation. Also, it should be noted that these predicted discharge values are calculated in Ethiopia while the observed one is recorded at Aswan. Several reasons can be given for the loss of precipitated water during its passage from Ethiopia to Aswan, including the effects of human habitation, of evaporation and of vegetation.

## 6.7 Summary and conclusions

Analysis of the observed and predicted rainfall and discharge over Ethiopia shows that the suggested rainfall model is applicable with a high accuracy for rainy months. This is because the model estimates the maximum precipitable water in the area under study due to the effect of water vapour content only, without taking into consideration clouds over the area which may give unpredictable rain. In general, it can be concluded that the

**Table 6.1** Observed and predicted values for the amount of rainfall over Ethiopia

| | J | F | M | A | M | J | J | A | S | O | N | D | Total |
|---|---|---|---|---|---|---|---|---|---|---|---|---|---|
| Observed | 20 | 30 | 62 | 87 | 61 | 105 | 290 | 272 | 170 | 40 | 19 | 18 | 1174 |
| Predicted | 244 | 37.1 | 59.5 | 73 | 63.4 | 106.4 | 152.6 | 155.7 | 106.9 | 43.2 | 24.8 | 16.4 | 863.4 |
| Corrected predicted | 25.08 | 33.03 | 58.47 | 78.35 | 57.58 | 92.66 | 239.73 | 225.42 | 144.33 | 40.98 | 24.29 | 23.49 | 1043.4 |
| Difference between observed and corrected predicted | −5.08 | −3.03 | 3.53 | 8.65 | 3.42 | 12.34 | 50.27 | 46.58 | 25.67 | −0.98 | −5.29 | −5.49 | 130.6 |
| Percentage error | 25 | 10 | 5.7 | 9.9 | 5.6 | 11.7 | 17.3 | 17.2 | 15.1 | 2.4 | 27.8 | 30.5 | 11.1 |

**Table 6.2** Observed and predicted values for fresh water discharge for Aswan and Ethiopia

| | J | F | M | A | M | J | J | A | S | O | N | D | Total |
|---|---|---|---|---|---|---|---|---|---|---|---|---|---|
| Observed | 4 | 3 | 2.5 | 2.4 | 2.5 | 3.5 | 7.6 | 19.6 | 20.7 | 13.5 | 7.5 | 5.3 | 92.1 |
| Predicted | 4.4 | 4.4 | 2.1 | 3.8 | 4.7 | 5.8 | 9.1 | 17.2 | 19.5 | 16.8 | 9.5 | 6.9 | 104.2 |
| Corrected predicted | 5.49 | 4.62 | 4.19 | 4.10 | 4.19 | 5.06 | 8.62 | 19.05 | 20.00 | 13.75 | 8.53 | 6.62 | 104.2 |
| Difference between observed and corrected predicted | -1.49 | -1.62 | -1.69 | -1.70 | -1.69 | -1.56 | -1.02 | 0.55 | 0.7 | -0.25 | -1.03 | -1.32 | -12.1 |
| Percentage error | 37.3 | 54 | 67.6 | 70.8 | 67.6 | 44.6 | 13.4 | 2.8 | 3.4 | 1.9 | 13.7 | 24.9 | 13.1 |

prediction model can be applied with a maximum of 15% error, which is considered to be satisfactory for such a prediction.

In general, the model suggested yields slightly higher values than those observed at Aswan. This can be justified due to the loss of floodwater during its passage from Ethiopia to Aswan. If the amount of water loss is included in the model at a later stage, the model may give better results for the whole year. In any case, the model proves that meteorological information could provide reliable evidence of expected floods.

# References

Brunt, D. (1952) *Physical and Dynamical Meteorology*, Cambridge University Press, Cambridge.
Haltimer, G. J. and Mortin, F. L. (1957) *Dynamical and Physical Meteorology*, McGraw-Hill, New York.

# 7 THE ENSO PHENOMENON AND ITS IMPACT ON THE NILE'S HYDROLOGY

Bayoumi B. Attia and Abdul Badie Abulhoda
*Water Research Center, Ministry of Public Works and Egypt Water Resources, Egypt*

## 7.1 Introduction

The Nile is the only major source of water for Egypt, and it was imperative that the country should try to regulate the river, preserve its water and make the most efficient use of this natural source of life. Forecasting the Nile flows at Aswan is of considerable importance. Good forecasts for several months ahead mean better control of Nile waters, especially during catastrophic situations.

Over the past two decades, various efforts have been made to develop mathematical models for forecasting the Nile's streamflow (Curry and Bras, 1978; Mobarek *et al.*, 1979; Salem *et al.*, 1983; Ministry of Irrigation, 1980). Nonetheless, it was consistently found that the performance of these models was only adequate during the low-inflow period (November-July) while in the high-inflow one (August-October) the quality of forecasting was unsatisfactory. Such results are consistent with the physical nature of a problem that prevents prediction of river flow before rainfall. This is a serious disadvantage, as the months when forecasts are urgently needed are those of the flood season.

On the other hand, considerable research has recently focused on the possible relationships that may exist between a climate event termed the El Niño/Southern Oscillation (ENSO) phenomenon, and other climate anomalies worldwide (Nicholls, 1985; Ropelewski and Halpert, 1987). These relationships have been referred to as 'teleconnections'. Recent studies (Ogallo, 1985; Ministry of Irrigation, 1987) illustrate the negative relationships between rainfall over many parts of east Africa and the ENSO event. However, the relationship between ENSO and rainfall over that part of the Ethiopian plateau which contributes to Nile inflows has not been thoroughly investigated.

This chapter is an attempt to provide an insight into the possible teleconnection between ENSO and Nile flows via statistical analysis. The well-defined life cycle of the ENSO event provides promise for forecasting future Nile floods once the validity of this teleconnection is established.

## 7.2 El Niño/Southern Oscillation

The El Niño phenomenon is the occasional flow of warm surface water from the western equatorial part of the Pacific to the eastern equatorial region (Nicholls, 1985). During El Niño events the sea surface temperature (SST) is exceptionally high (several degrees above normal) over much of the equatorial eastern and central Pacific.

In the late 1960s it became apparent that the year-to-year variations in SSTs (El Niño) were closely linked to the Southern Oscillation, an out-of-phase relationship between atmospheric pressures over the south-eastern Pacific and the Indian Ocean. Figure 7.1 shows the relationship between SST on the equatorial Pacific and the Southern Oscillation. The two events are now more broadly referred to as the El Niño/Southern Oscillation (ENSO) phenomenon.

The development of ENSO can be divided into four phases:

1. *Precursors*: There are stronger than usual surface easterly winds in the western equatorial Pacific in the months before a strong ENSO. Associated with this are lower than average pressures over the Indonesian region and above-average ones in the south-eastern Pacific. The SST is slightly warmer than average in the west and colder in the east.

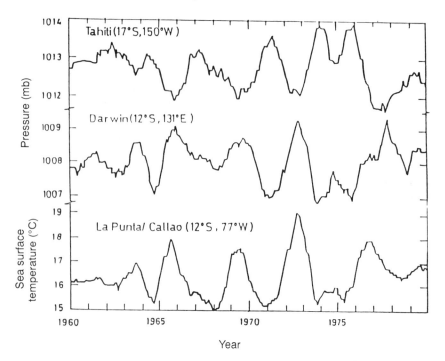

**Figure 7.1**   *SST on the equatorial Pacific and Southern Oscillation (twelve-month running means)*

2. *Onset*: In December the warm SST anomalies disappear from the western equatorial Pacific and the surface wind anomalies suddenly switch from easterly to westerly. Warm SST anomalies appear in the central Pacific.
3. *Growth*: Above-average SSTs of the South American coast appear in February or March, increasing up to June and the strong westerly wind moves east. Pressure is much higher than average at Darwin (Indian Ocean) and much lower at Tahiti (Pacific).
4. *Decay*: West of the dateline, westerly wind anomalies start to weaken from December onwards and SSTs increase. There is a second warming at the South American coast, peaking early in January followed by a rapid decrease to colder than average SSTs. The pressures at Darwin and Tahiti return to normal.

## 7.3 Hydrology of the Nile Basin

The Nile, 6500 km in length, is the second longest river in the world. Over its great basin (10% of the African continent) there are large variations in climate, topography, geography and hydrology. There are two primary sources of the main Nile channel: the White Nile and the Blue Nile. Combining with these sources to augment the Nile's flows are three other major tributaries: the Bahr El Ghazal, the Sobat and the Atbara (Figure 7.2).

The White Nile is the most distant source, emerging from Lake Victoria on the equatorial lakes plateau. From the equatorial lakes the river flows northward to the Sudd region. The average outflow from the Sudd is about 15 milliard $m^3$/yr. Before reaching the confluence with the Blue Nile, the White Nile is joined by the Bahr El Ghazal and the Sobat. The Bahr El Ghazal is a very swampy river, located in the region of south-western Sudd, that adds an average of 0.5 milliard $m^3$/yr to the White Nile. The Sobat, originating in the Ethiopian plateau, adds an average of 13.5 milliard $m^3$/yr to the White Nile. Thus the total amount of water provided by the White Nile at Malakal is, on average, 29 milliard $m^3$/yr and distribution is relatively uniform.

The other major source of the main Nile is the Blue Nile. Emerging from Lake Tana, in the Ethiopian plateau, the Blue Nile is characterized by its torrential nature, carrying violent floods during the months of August to October. The annual yield of the Blue Nile at Khartoum is approximately 54 milliard $m^3$/yr. Like the Blue Nile, the Atbara is torrential in nature. Its average annual yield is approximately 12 milliard $m^3$. From the junction of the Nile with the Atbara, the Nile flows through a desert climate, reducing the amount of water arriving at Aswan to an estimated average of 84 milliard $m^3$/yr.

It is worth noting that almost 85% of Nile flows at Aswan is the water originating in the Ethiopian plateau. Most of this arrives in Egypt during the months of August to October.

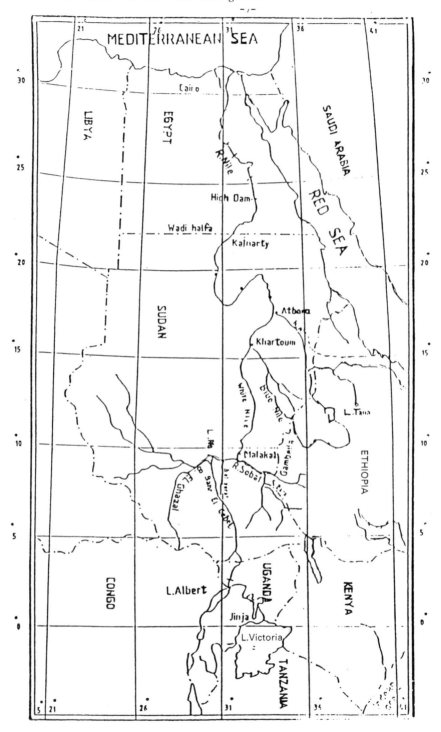

**Figure 7.2**   *Nile Basin and key streamflow-measuring stations*

## 7.4 Nile flows and ENSO teleconnections

The purpose of this section is to consider a possible relation between ENSO events and the flows of the Nile. Data for monthly streamflow records of four key stations are available, but as a mixture of natural and regulated flows. These stations are Malakal (White Nile), Khartoum (Blue Nile), Atbara (River Atbara) and Aswan, as illustrated in Figure 7.2.

The data are for the period 1914–1988 and were measured either before or after the construction of existing dams (Gabal El Aulia, Kashem El Gerba, Roseries, and Sennar). A procedure was undertaken to obtain a batch of completely normalized data, as illustrated in Figures 7.3–7.5. It can be assumed with reasonable accuracy that these data are normally distributed.

The 75 years covered by this study spans fourteen ENSO events. Specifically, the following years are the most significant during the period 1914–1988 (Nicholls, 1985; Ropelewski and Halpert, 1987; Yoshino and Yasanari, 1985):

1918, 1925, 1930, 1939, 1941, 1951, 1953, 1957, 1963, 1965, 1969, 1972, 1976, 1982

Each ENSO event is different, and thus it is difficult to obtain a precise quantitative measurement of its magnitude. Only two categories of ENSO

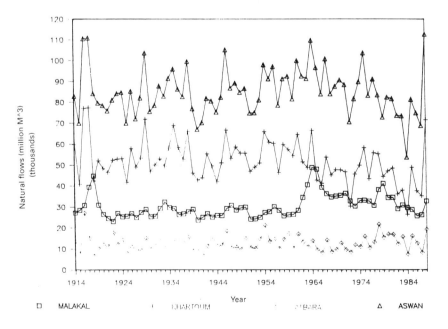

**Figure 7.3** *Natural flows at key stations on the Nile during a calendar year*

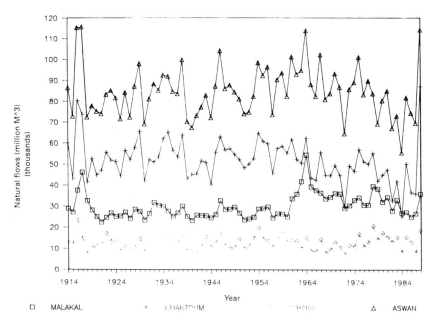

**Figure 7.4**    *Natural flows at key stations on the Nile during an hydraulic year*

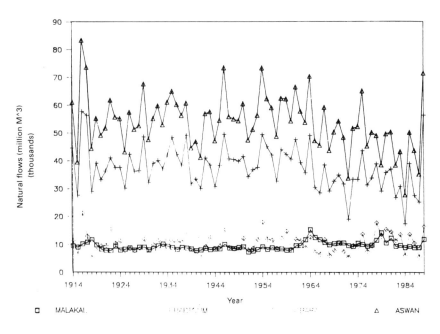

**Figure 7.5**    *Natural flows at key stations on the Nile during a flood period*

episodes are considered. Each year is regarded as either an ENSO event or not. Due to these data restrictions a conventional correlation analysis between Nile flow and ENSO is inappropriate. The adopted method of analysis is based on the biserial correlation (Arkin and Cotton, 1948), which is outlined below.

When there are only two categories of one attribute and a number of normally distributed classifications for the other, the biserial correlation coefficient may be computed as:

$$r = (\overline{X}_p - \overline{X}_q)pq/0.3989\sigma h$$

where

$\overline{X}_p$ = the mean value of the $p$ category,
$\overline{X}_q$ = the mean value of the $q$ category,
$p$ = percentage of cases in the $p$ category,
$q$ = percentage of cases in the $q$ category,
$\sigma$ = standard deviation of the combined categories,
$h$ = height of ordinate of the normal curve at a distance from the mean, including $(p - q)/\sigma$ of the area of the curve.

For each station the biserial correlation is computed between the ENSO event and three attributes, namely: total inflow during the calender year January–December, total inflow during the hydraulic year August–October, and total inflow during the flood season August–September.

The results are summarized in Table 7.1, from which we can draw the following observations:

1. There exists a significant teleconnection between an ENSO event and the flow in the Blue Nile (Khartoum station) and the Atbara (Atbara station) during the flood season, calendar year and hydraulic year. This means that, in years when there is an ENSO event, one may expect a low or below-average flood in both the Blue Nile and the Atbara.
2. The relationship between ENSO events and the flow of the White Nile (Malakal station) is too weak to be considered as a teleconnection. This result was not surprising, as the rainfall regimes of the equatorial lakes are different from those of the Ethiopian pleateau.

**Table 7.1** Biserial coefficient of correlation between ENSO and natural flows at different stations

|  | *Atbara* | *Khartoum* | *Malakal* | *Aswan* |
|---|---|---|---|---|
| Flood period | −0.58 | −0.60 | −0.02 | −0.56 |
| Hydraulic year | −0.64 | −0.54 | −0.09 | −0.48 |
| Calendar year | −0.56 | −0.53 | +0.19 | −0.35 |

3. Inflow at Aswan is significantly associated with ENSO events during the flood season. Nonetheless, the total inflow at Aswan during the entire calendar year is weakly linked to ENSO events. One explanation is that the major contribution of the Aswan inflow during the flood season is due to the Blue Nile and the Atbara, which show a significant teleconnection with ENSO events. However, for the rest of the year White Nile flows compensate for the lowest discharges in both the Blue Nile and the Atbara.

## 7.5 Conclusion

The link between the climatic phenomenon called the El Niño/Southern Oscillation (ENSO) and the Nile flows has been studied. It has been verified statistically that ENSO episodes are negatively teleconnected with the floods of the Blue Nile and the Atbara, and consequently with Aswan inflow during the flood season. Due to the well-defined life cycle of ENSO episodes, low or below-average floods at Aswan may be expected several months ahead. However, it should be emphasized that only the sufficiency (and not the necessity) part of this teleconnection is verified. That is, low floods may also occur in non-ENSO years.

## References

Arkin, H. and Cotton, R. (1948) *An Outline of Statistical Methods*, 4th edn, W. Heffer, Cambridge.

Curry, K. and Bras, R. L. (1978) *Theory and Applications of the Multivariate Broken Line, Disaggregation and Monthly Autoregressive Streamflow Generator to the Nile River*, Technology Adaptation Program, MIT, Cambridge, MA.

Ministry of Irrigation (1980) Master Water Plan Project, *Hydrological Simulation of Lake Nasser*, Technical Report 14 (UNDP-EGY/73/024), Cairo.

Ministry of Irrigation (1987) *Rehabilitation and Improvement of Water Delivery Systems in Old Land*, National Irrigation Improvement Project, UNDP Project No. EGY/85/012, First Interim Report, February, Cairo.

Mobarek, I. E., Salem, M. H. and Dorrah, H. T. (1979). 'Hydrological studies on the River Nile, (I) Forecasting, (II) Stochastic modeling', *Research Report*, Cairo University/MIT Technological Planning Program, Cairo.

Nicholls, N. (1985) 'The El Niño/Southern Oscillation phenomenon', Workshop on the Societal Impacts Associated with the 1982–83 Worldwide Climate Anomalies, Lugano, Switzerland, November.

Ogallo, L. (1985) 'Impacts of the 1982–83 ENSO event on eastern and southern Africa', Workshop on the Societal Impacts Associated with the

1982–83 Worldwide Climate Anomalies, Lugano, Switzerland, November.

Ropelewski, C. T. and Halpert, M. S. (1987) 'Global and regional scale precipitation patterns associated with the El Niño/Southern Oscillation', *American Meteorological Society, Monthly Weather Review*, **115**, No. 8, August.

Salem, M. H., Dorrah, H. T. and Alawi, M. M. (1983) 'Multivariate auto-regressive least-squares forecasting model for the River Nile', Conference on Water Resources Development in Egypt, Cairo.

Yoshino, M. M. and Yasanari, T. (1985) 'Climatic anomalies of El Niño and anti-El Niño years and their socio-economic impact in Japan', Workshop on the Societal Impacts Associated with the 1982–83 Worldwide Climate Anomalies, Lugano, Switzerland, November.

# 8 A POWER SPECTRUM FOR THE NILE's NATURAL FLOW TIME SERIES

Abdel-Azim M. Negm, *Zagazig University, Egypt*
Bayoumi B. Attia, *Water Research Center, Ministry of Public Works and Water Resources, Egypt* and
Talaat M. Owais, *Zagazig University, Egypt*

## 8.1 Introduction

This chapter discusses the problem of estimating both discrete (line) and continuous spectrum values for a hydrologic time series. In the case of a discrete spectrum the least-squares technique is used to estimate Fourier coefficients. In turn, the coefficients obtained are used to estimate the periodogram or the discrete spectrum ordinates, while in the case of the continuous (power) spectrum, the relation between the continuous spectrum and autocorrelation of a continuous function (time series) is used. The unbiased estimate of the continuous (power) spectrum is obtained through the use of a smoothing filter and two smoothing functions (filters) are discussed. Finally, the two methods are applied not only to identify the structure but also to obtain the hidden periodicities (if any) of the Nile's natural flow time series at Aswan.

## 8.2 Discrete spectrum analysis

Consider a trend and jump-free series $X_t$ ($t = 1, \ldots, N$) which may satisfy a linear additive model of the form (Adamowski, 1971)

$$X_t = p + Y_t \tag{8.1}$$

where $p$ is a periodic component (it may be simple or complex) and $Y_t$ is a stochastic component. Let the mean of the series $X_t$ be $\mu$. Then the series can be represented by the spectrum model

$$X_t - \mu = X_{est} + \epsilon_t \tag{8.2a}$$

where $\epsilon_t$ is a random error term and $X_{est}$ is given as

$$X_{est} = \sum_{i=1}^{h} C_i \sin(2\pi ti/N + \phi_i) \quad i = 1,2,3, \ldots ,N \qquad (8.2b)$$

$h$ = the total number of the harmonics and
$h$ = $N/2$ if $N$ is even or $(N-1)/2$ if $N$ is odd,
$t$ denotes time ($t = 1,2, \ldots ,N$),
$N$ is the total number of observations of series points,
$i/N$ is the frequencies of occurrence of harmonics,
$N/i$ is the wavelengths of the harmonics,
$C_i$ and $\phi_i$ are the amplitudes and phase angles, respectively.

The phase angle $\phi_i$ signifies the horizontal alignments of the sine curves with respect to the origin.

The parameters $C_i$ and $\phi_i$ can be estimated by another two parameters and are in the equivalent form of the model given by (Negm, 1989; Adamowski, 1971; Anderson and Lomb, 1980; Kottegoda, 1980):

$$X_{est} = \sum_{i=1}^{h} [\beta_i \sin (2\Pi ti/N) + \alpha_i \cos (2\Pi ti /N)] \qquad (8.3)$$

where $\beta_i$ and $\alpha_i$ are the new parameters (Fourier coefficients) and should be estimated.

Let $\hat{\mu}$, $\hat{\alpha}$, and $\beta$ be the least squares estimates of the parameters $\mu$, $\alpha$ and $\beta$ as given by:

$$\hat{\mu} = \frac{1}{N} \sum_{t=1}^{N} X_t \qquad (8.4)$$

$$\hat{\alpha}_i = \frac{2}{N} \sum_{t=1}^{N} X_t \cos (2\pi ti/N) \qquad (8.5)$$

$$\beta_i = \frac{2}{N} \sum_{t=1}^{N} X_t \sin (2\pi ti/N) \qquad (8.6)$$

and $i = 1,2,3, \ldots ,h$.

The parameters $C_i$ and $\phi_i$ can now be expressed in terms of $\hat{\alpha}_i$ and $\beta_i$ as given by:

$$C_i^2 = \alpha_i^2 + \beta_i^2 \qquad (8.7)$$

$$\phi_i = Tan^{-1} \beta_i/\alpha_i \qquad (8.8)$$

The unbiased estimates of the periodogram ordinates at the different frequencies $P$ $(i/N)$ can be estimated using the same parameters of equations (8.5) and (8.6) as given by

$$P(i/N) = C_i^2/2 = (\hat{\alpha}_i^2 + \beta_i^2)/2 \tag{8.9}$$

To test the significance of the periodogram ordinates corresponding to a periodicity given by the inverse of its frequency, a Fisher test is used. This uses the following ratio (Blackmann and Tukey, 1958; Yevjevich, 1972):

$$g_i = \frac{C_i^2}{2S_x^2 - \sum_{j=1}^{i-1} C_j^2} \tag{8.10}$$

where $C_i^2$ is the maximum periodogram ordinate and $S_x^2$ is the estimate of the variance of the series $X$.

The largest harmonic is statistically significant if the ratio $g_1$ exceeds the ratio $g$ given by the following equation for a probability level $P_f$ for $h$ harmonics:

$$P_f = \sum_{j=1}^{k} (-1)^{j-1} \frac{h! \, (1 - jg)^{h-1}}{(h - j)! \, (j - i)! \, (i - 1)! \, j!} \tag{8.11}$$

where $k$ is the greatest integer less than $1/g$. The first term on the right-hand side of equation (8.11) usually gives a sufficient approximation for $g$-critical (Fisher, 1929; Yevjevich, 1972).

The values of equation (8.10) are biased, since they produce more significant harmonics due to the subtraction of the greatest harmonics from the data. As a result of this subtraction, the variance becomes smaller and hence the $g$-critical value larger as the denominator of equation (8.10) becomes smaller. Therefore this method is an approximate procedure, but it satisfies the needs of the analysis of the hydrologic time series which may be periodic, regardless of the built-in bias.

## 8.3 Continuous spectrum analysis

As discussed in the previous section, the discrete spectrum has difficulties and thus there is a deficiency. As a result, the results obtained by applying the discrete spectrum may be confusing for the analyst. By applying the continuous spectrum to the hydrologic time series data the deficiencies of the discrete spectrum can be avoided. The continuous spectrum is sometimes called a power spectrum or variance density spectrum (VDS). Here the three terms will be used synonymously.

The power spectrum estimate $PS(f)$ of a continuous series $X_t$ ($t = 1, N$ and $N \ni$) is related to its corresponding continuous autocorrelation function by the relation given by

$$PS(f) = 2 \left[ \rho_0 + 2 \sum_{k=1}^{k_{max}} \rho(k) \cos 2\pi f k \right] \qquad (8.12)$$

where $\rho(k)$ is the population autocorrelation, which is estimated by the sample autocorrelation $r(k)$, and $\rho_0$ is the first autocorrelation which should be replaced by unity (Blackmann and Tukey, 1958; Yevjevich, 1972).

The practical range for $f$ is $0 \subseteq f \subseteq 0.5$, because $f_{min} = 1/N$, and as $N$ tends to infinity, $f$ tends to zero, and the largest frequency $f_{max} = 1/(2\Delta t) = 0.5$. If the series is finite, the above limit $f_{min}$ will depend on the sample size. To avoid this dependence of the spectrum on the series length, $f_{min}$ is taken as 0 and the small range $0 \subseteq f \subseteq 1/N$ is estimated similarly for the other sequences.

*8.3.1 Estimating the unbiased power spectrum*

To estimate a continuous spectrum in the range $0 \subseteq f \subseteq 0.5$, the range is divided into intervals of $\Delta f$. The total number of frequencies of the power spectrum to be estimated is then given by

$$m = 1 + 0.5/\Delta f \qquad (8.13)$$

The estimates obtained by equation (8.12) are biased because they are somewhat greater than the expected or population variance density spectrum. In other words, let (Negm, 1989; Blackmann and Tukey, 1958; Yevjevich, 1972)

$$E[P\hat{S}(f) - PS(f)] = \epsilon \qquad (8.14a)$$

$$\text{Var } P\hat{S}(f) = E\{E[P\hat{S}(f) - PS(f)]\}^2 = \psi \qquad (8.14b)$$

Then for the estimates to be unbiased and efficient the right-hand side of equation (14a) should be zero and the right-hand side of equation (8.14b) should be minimum. This can be achieved by making a modification of the spectrum obtained by equation (8.12) for the selected range of $f$ either by smoothing the $r(k)$ or by smoothing the computed adjacent discrete spectrum estimates for a limited number of frequencies.

*8.3.2 Modification of the power spectrum estimates*

The power spectrum estimates can be modified by smoothing either the correlogram or the power spectrum ordinates using a smoothing function. Assuming that this function in the lag domain is denoted by $D(k)$, then when smoothing the finite correlogram with $k_{max} = N/2$ for a series of length $N$, by a smoothing function $D(k)$, equation (8.11) becomes

$$P\hat{S}(f) = 2 \left[ r_o + 2 \sum_{k=1}^{k_{max}} D(k) \, r(k) \cos 2\Pi fk \right] \qquad (8.15)$$

Two smoothing functions (filters) will be considered here, the Hanning and the Tukey, as given by:

$$D(k) = (1 + \cos 2\pi fk)/2 \qquad (8.16)$$

$$D(k) = 1 - 2a + 2a \cos (\pi k/h) \quad a = 0.23 \qquad (8.17)$$

## 8.4 Confidence and tolerance limits

The estimates of the power spectra (variance density spectrum) ordinates for a normal and independent series for the range of frequencies $0 \subseteq f \subseteq 0.5$ are independently distributed as a $\chi^2$-variable, which is a good approximation, for finite $N$. Therefore to test the significance of a particular ordinate (peak) the value of the power spectrum ordinate is compared to that of the $\chi^2$ divided by an equivalent degree of freedom $v$, or the square of the former is compared to the square of $\chi^2/v$, where $v$ is given by $8N/3h$ to $2N/h$ for the filter (smoothing function) of Hanning and Tukey (Negm, 1989; Yevjevich, 1972). The value $\chi^2/v$ is called the upper tolerance limit, while the lower tolerance limit is denoted $(1 - X^2)/v$. The upper confidence interval is obtained by multiplying the upper tolerance limit by the power spectra ordinate at each particular frequency.

## 8.5 Analysis of results

The Nile's natural flow time series at Aswan are analysed by using both discrete and power spectrum analysis. The results show that the series have a complex periodic variation. The set of statistically significant periodicities investigated by discrete spectrum analysis and tested by the Fisher test as well as that obtained through the use of the continuous spectrum analysis is listed in Table 8.1. This table shows that the largest

**Table 8.1** Periodicities within the first third harmonics of the Aswan station series using discrete and continuous spectra

| Month | Discrete spectrum | Power spectrum Hanning | Tukey | Month | Discrete spectrum | Power spectrum Hanning | Tukey |
|---|---|---|---|---|---|---|---|
| Jan. | 14.5 | 14.5 | 14.5 | Aug. | 58.0 | 7.3 | 7.3 |
|  | 11.6 | 29.0 | 19.3 |  | 7.3 | 14.5 | 19.3 |
|  | 6.1 | 7.7 | 10.5 |  | 19.3 |  |  |
|  | 7.7 | 19.3 | 7.7 |  | 11.6 |  |  |
| Feb. | 14.5 | 29.0 | 29.0 | Sep. | 7.3 | 14.5 | 14.5 |
|  | 38.7 | 19.3 | 19.3 |  | 16.6 | 7.3 | 7.3 |
|  | 7.7 | 12.9 | 12.9 |  | 6.4 | 19.3 | 6.4 |
|  | 6.4 | 7.7 | 7.7 |  | 8.9 | 29.0 | 19.3 |
| March | 16.6 | 7.7 | 38.7 | Oct. | 58.0 | 14.5 | 14.5 |
|  | 7.7 | 19.3 | 16.6 |  | 14.5 | 6.4 | 19.3 |
|  | 23.2 | 29.0 | 12.9 |  | 6.4 | 19.3 | 6.4 |
|  | 11.6 | 16.6 | 7.7 |  | 7.3 | 7.7 | 7.3 |
| April | 16.6 | 19.3 | 16.6 | Nov. | 38.0 | 38.0 | 19.3 |
|  | 7.7 | 7.7 | 7.7 |  | 19.3 | 6.4 | 6.4 |
|  | 23.2 | 12.9 | 12.9 |  | 6.4 | 19.3 | 14.5 |
|  | 9.7 |  | 9.7 |  | 10.5 | 12.9 |  |
| May | 7.7 | 7.7 | 38.0 | Dec. | 14.5 | 19.3 | 19.3 |
|  | 16.6 | 16.6 | 16.6 |  | 6.1 | 6.1 | 14.5 |
|  | 9.7 | 9.7 | 23.2 |  | 11.6 | 12.9 | 6.1 |
|  | 23.2 | 12.9 | 7.7 |  | 7.3 | 7.7 |  |
|  |  | 5.5 | 12.9 |  |  |  |  |
| June | 7.0 | 7.7 | 7.7 | Total | 19.3 | 14.5 | 19.3 |
|  | 9.7 | 9.7 | 9.7 |  | 7.3 | 19.3 | 14.5 |
|  | 23.2 | 12.9 | 23.2 |  | 6.4 | 7.7 | 10.5 |
|  | 16.6 | 19.3 | 16.6 |  | 11.6 |  |  |
| July | 8.3 | 8.3 | 38.0 |  |  |  |  |
|  | 19.3 | 16.5 | 8.3 |  |  |  |  |
|  | 9.7 | 6.4 | 23.2 |  |  |  |  |
|  | 6.4 | 10.5 | 9.7 |  |  |  |  |

periodicity in the flood period is 58 years, while periodicities of 14.5, 16 and 38 years are dominant in low-flow periods. The annual flow series has the largest periodicity of 19.3 years in the case of discrete spectrum analysis, which appeared again when the power spectrum with a Tukey smoothing filter was used. A Hanning smoothing filter gives the first significant periodicity as 14.5 years and the second as 19.3 years.

The results of discrete spectrum analysis are used to compute the contributions of the different components to the total explained variance. The variance of each harmonic can be calculated by dividing the ordinate of this harmonic by the sum of all ordinates. The series is found to consist of three components: trend, periodic and stochastic. Figure 8.1 shows that the contribution of the stochastic component to the total explained variance in both the series of August and October is dominant while that of the trend is very small. Figure 8.2 shows the results of analysing the subseries of equal length of 50 years, and that the main periodicities (the first and the second in two subseries) are not the same. Figure 8.3 gives the explained variance of each of the two periodicities. It is obvious that the explained variance is not the same, even in the case of two equal periodicities. The results of analysing three subseries of 40 years' length are similar to those of the 50 years' subseries.

## 8.6 Conclusions

1. The explained variance by the stochastic component is dominant for the flood months, while that of the trend is small compared to that of other months.
2. The small periodicities do not cover the full length of the series. This leads the analyst to make a study of both the full-length series and the subseries.

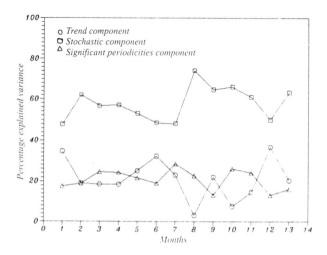

**Figure 8.1** *Contributions of the various components to the total explained variance in the Aswan station series*

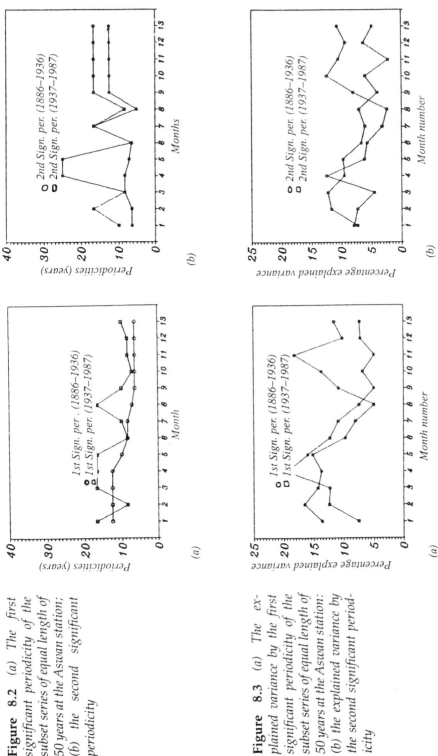

**Figure 8.2** (a) The first significant periodicity of the subset series of equal length of 50 years at the Aswan station; (b) the second significant periodicity

**Figure 8.3** (a) The explained variance by the first significant periodicity of the subset series of equal length of 50 years at the Aswan station: (b) the explained variance by the second significant periodicity

3. The trend component is not a perfect trend but can be represented either by a constant trend or by a short-period one.
4. All the significant periodicities of the full-length trend cannot be used in forecasting, since some of them occur only in a particular period.
5. The results obtained may explain the deficiency of applying statistical methods for long-term forecasting.

# References

Adamowski, K. (1971) 'Spectral density of a river time series', *J. Hydrol.*, No. 14, 43–52.

Andel, J. and Balek, J. (1971) 'Analysis of periodicity in hydrological sequences', *J. Hydrol.*, No. 14, 66–82.

Anderson, A. P. and Lomb, N. R. (1980) 'Yet another look at two classical time series', Research paper, ITSM.

Blackmann, R. B. and Tukey, J. W. (1958) *The Measurements of Power Spectra*, Dover, New York.

Bloomfield, P. (1976) *Fourier Analysis of Time Series: An Introduction*, John Wiley, New York.

Fisher, R. A. (1929) 'Tests of significance in harmonic analysis', *Proc. R. Soc. London, Ser. A*.

Julian, P. R. (1967) 'Variance spectrum analysis', *Water Resour. Res.*, No. 3, 831–845.

Kottegoda, N. T. (1980) *Stochastic Water Resources Technology*, Macmillan, London.

Michelson, A. A. (1913) 'Determination of periodicities by the harmonic analyzer with an application to the sunspot cycle', *Astrophys. J.*, **38**.

Negm, A. Z. M. (1989) *Study of Surface Water Hydrology With Application to River Nile: Natural Nile Flows Forecasting Via Spectrum Analysis*, MSc. Thesis, Faculty of Engineering, Ain Shams University.

Schuster, A. (1948) 'On investigation of hidden periodicities', *Terr. Magn.*, No. 3, 13–41.

Yevjevich, V. (1972) *Stochastic Process in Hydrology*, Water Resources Publications, Fort Collins, Colorado.

# Part 2

# Impact of Climatic Changes

# 9 ASSESSMENT OF THE HISTORICAL CHANGES OF AQUATIC ENVIRONMENTS UNDER CLIMATIC IMPACTS

J. Balek
*UNEP, Nairobi*

## 9.1 Introduction

The earliest stages of man's existence in Africa were influenced by the fluctuation of wet and dry periods, the so-called interpluvials and pluvials. Wayland (1934) identified the main periods of increased rainfall separated by periods of drier climate at Lake Victoria. Nilsson (1932), together with Leakey, found two wet post-Pleistocene phases called Gamblian and Nakuran, and a Kamasian and Kageran period were identified in the Middle and early Pleistocene, respectively.

It is not surprising that a rich cultural history was correlated by Leakey (1964) with the periods of pluvials and faunal development. The reconstruction of past trends in ecological development has become a significant part of Quaternary archaeology in Africa.

The regions of similar summer and winter rainfall probably remained the same throughout the Quaternary. Instead of changes in the ecological systems there was some shifting of the vegetational and faunal zones and ecological boundaries; these shifts were conceivably extensive. For instance, during so-called Late Acheulian and Aterian times the Sahara desert was invaded by Mediterranean flora down to the southern limits. This required at least 700 mm of annual rainfall and more temperate conditions in areas which today are deserts.

## 9.2 Climate changes indicated by groundwater recharge

By using radioactive dating, the age of groundwater in the Nubian sandstones was determined as between 30 000 and 40 000 years. It is commonly agreed that elsewhere in the Sahara the significant groundwater recharge occurred earlier than 20 000 BP and between 10 000 and 20 000 BP, while the recharge concepts for the period between 10 000 and 20 000 BP are contradictory, and interpretations range from arid to humid climate. Also, the present dryness in the regions is compared with the previous extended dryness about 20 000 BP during the last glacial maximum. It was also concluded by Chaline (1985) that the Sahel was moister during the interglacials and during less cold stages of the last glacial than during the glacials (Laberyie, 1985). The present situation can be considered as an interglacial period, but small-scale glacier thawing in Europe between 1960 and 1980 cannot be correlated with the dry conditions in the Sahel.

For central Europe, Geyh and Backhaus (1978) concluded that during the glacial period the earth's surface was sealed by permafrost and groundwater recharge was interrupted. The groundwater recharge probably occurred during the so-called Weichselian interstadials up to 20 000 BP and after 12 000–10 000 BP the recharge was large enough to keep the hydrostatic pressure high (Balek, 1989).

Even if it is expected that the pluvials correspond to the glacial periods in the temperate regions, such a coincidence in the recharge processes in European and African sedimentary structures cannot be given a genetically common and simple explanation. Nevertheless, the magnitude of the variability of the recharge processes clearly indicates the size of the climatic impacts which have affected the history of the formation of groundwater resources.

The aridity of an area depends not only upon the climatic conditions but also upon geological and geomorphological development. These can be of considerable importance, particularly when they are of a late geological age. For instance, geomorphological uplifts in many parts of Africa may have resulted in the destabilization of the climate for a prolonged period. According to Dixey (1962), a typical trace of such a phenomenon can be a lack of harmony between the observed features of the soils, water supply and current climate.

While groundwater age and morphological features can provide an integrated point of view of the climate variability in Europe and Africa, changes in the ecosystems and fauna and flora can be traced on the rock paintings in the northern Sahara. Pictures of hunters and game found today much farther south can be dated relatively easily, and indicate that during the last 40 000 years the Sahara experienced at least five humid and arid periods — not because the fauna and flora had been changed under man's impact: on the contrary, it was fauna, flora and man which had to adapt themselves to the climate changes.

## 9.3 Climate changes as indicated by water-level fluctuation

At present, the observation network of African surface waters is far from being developed: on the contrary, in some regions it has been deteriorating. Nevertheless, even if hydrographic observations had been available from the beginning of the century it is doubtful that convincing methods can be applied to prove the impact of climate on a changing hydrological regime.

However, particularly in Africa (but to some extent also on the other continents), very convincing data are available which can be used to analyse the relation between the fluctuation of climate features and water-related phenomena. For instance, when historically long records of Nile levels became available, not only hydrologists (Andel *et al.*, 1971) but historians, archeologists and Egyptologists (Verner, 1972) became concerned with the fluctuation of the hydrological regime in the Nile Basin and with related phenomena. Attention has been focused on the Nile not only because the oldest known hydrological records are from that river but also the history of its water resources development is among the oldest in the world. Herodotus, who came to Egypt in the fifth century BP, found the oldest Nile dam built south of Memphis under the rule of Pharaoh Menes in about 3000 BP. At present, the oldest dam in the world is in the Wadi el Garawi, only 25 km from Cairo. It was built between 2950 and 2750 BP. The hydrological regime in the wadi was probably much more favourable for the construction of such a dam. Also, in the Middle Kingdom (2160–1788 BP) several artificial reservoirs were constructed, presumably by order of Amenemhet III. Again, such engineering works indicate more favourable climatic and hydrological conditions than today.

Distinct changes in the Nile Basin hydrology can be traced towards the end of the Middle Paleolithic. Then local precipitation in Egypt declined and the basin became attached to the Ethiopian highlands drainage system. Since then a regular regime of annual floods has been established in the Nile. These brought the first mud deposits, the so-called Basal or Selbian silts (Vingard, 1923), and it has become known as the siltation stage (Fairbridge, 1963). At the same time, widespread changes took place in the fauna and flora distribution, and people began to concentrate on the Nile's banks.

Sedimentation of Nile's mud with some erosional interruptions continued irregularly from the pre-Neolithic times up to the so-called Neolithic subpluvial, which lasted in Egypt from the early Neolithic until the late Old Kingdom (Verner, 1972). It is probable that, following the fading of the Neolithic wet phase, the hydrological regime continued to be influenced by an extended Makalian wet phase, and before 850 BP by the Nakurian wet phase in the Ethiopian highlands (Trigger, 1965).

Fairbridge (1962) applied geologically oriented findings and concluded that the Nile's bed became more or less stable and achieved its present form around 3000 BP. He also identified several outstanding periods (Table 9.1).

Butzer (1959) correlated the gradual disappearance of Neolithic savanna fauna and shifts of fauna with an overall fluctuation of the climate. So-called fauna breaks can be traced around 3600 BP, 2800–2600 BP and 2500–2300 BP. The last date coincides with the end of Neolithic subpluvial.

Butzer also recognized three stages in the shift of sand dunes on the Nile's west bank as another indication of the fluctuation of climate:

**Table 9.1**   Runoff variability (after Fairbridge, 1962)

| Higher runoff (BP) | Lower runoff (BP) |
| --- | --- |
| 8500 | 9500–9000 |
| 7000 | 7500 |
| 5500 | 4500 |

2350–500 BP, AD 1200–1450 and after AD 1700. By combining his findings, Butzer divided the period from the end of Neolithic subpluvial until the present into several hydrological stages (Table 9.2). Heinzelin (1967) correlated the state of the Nile's floods by the radioactive dating in selected localities of Wadi Halfa (Table 9.3). In the fragments of annals coming from the Old Kingdom (Gauther, 1914; Sethe, 1933) it can be seen that as early as the First Dynasty the rulers kept records of flood levels year by year, and fragments of records can be found for all five dynasties. They are, however, not suitable for analysis, first, because of their fragmentary nature and second, because their consistency with later Arabic records cannot be proved. In general, records expressed in cubits are relatively lower than the later records of Arabs. This can be partly explained by different units (Popper, 1951). The ancient Egyptian cubit was equal to 7 palms or 28 fingers and corresponded to 0.523 m. Jequier (1906) expected that the zero of the nilometers in ancient Egypt was very much different from the nilometers of Arabs, and that on the Old Kingdom's nilometers the observations were read not upwards but downwards.

**Table 9.2**   Hydrological regimes (after Butzer, 1959)

| | |
| --- | --- |
| 2350–500 BP | Dry climate, intensive aeolian deposits |
| 500 BP–AD 300 | Higher floods, increased sedimentation |
| AD 300–800 | Lower runoff, prevailing aeolian deposits, local drought |
| AD 800–1200 | Higher floods |
| AD 1200–1400 | Dry period, younger dunes |
| AD 1450–1700 | Higher floods, local drought |
| AD 1700 onwards | Younger dunes in Middle Egypt, drought |

Important observations of the Nile's water level from the Middle Kingdom come from Nubia. Unfortunately, these data are too fragmentary for statistical analysis. Twenty-seven fragments were discovered at Kumma, mostly from the period of Amenhotep III. Reisner *et al.* (1960) analysed the records and concluded that at that time the Nile flood level was some 7 m higher than at present. At Uronarti, Shalfak and Mirgissa, where all fortresses have staircases descending to the river, traces of bricks can be

**Table 9.3** Nile fluctuation (after Heinzelin, 1967)

| | |
|---|---|
| Predynastic period (Old Kingdom) | Low runoff |
| Middle Kingdom | High runoff |
| New Kingdom | General decrease of runoff |
| Meroitic period | Low runoff, sand deposits |
| Christian period | Higher floods |

found more than 7 m above the present level of the floods. Also, traces of mud can be found 6.5 m above the present highest level.

Fragments of records discovered at the end of the nineteenth century at Karnak originated from the Third Intermediate Period and the Late Period. They indicate that records of the Nile's water level have been maintained continuously. However, the most valuable records have been preserved only from the early stage of the Arabic period.

The basic set of those data was compiled and published by Prince Omar Toussun (1925). They originate from data gathered by Ibn al-Hidjazi and found in an unpublished manuscript *Nail ar-Ra" idmin an-Nil az-Za'id* (Bibliothèque Nationale, Arabic MS 2261). The missing data were supplemented by the statistics compiled by Ibn Taghri Birdi in *an-Nudjum az-Zahira* (Langles, 1810). The source of Ibn al-Hidjazi's data is not known. Popper (1951) concluded that the records of minima were taken close to the present date of 20 June while the maxima records disregard any date.

Andel *et al.* (1971) attempted to use both sets of data to analyse periodical components in the river regime and relate them either to climate fluctuation or to extraterrestrial impacts. However, the quality of those records cannot be compared with that of water level measurements at present.

Toussun attempted to translate historical measurements in cubits and fingers into the metric system (Popper, 1951). Transformed and published data contain many inaccuracies and numerous errors can be expected in the original records compiled by the Arab chroniclers.

The following must be kept in mind when analysing and translating the data expressed in cubits and fingers:

1. Two different kinds of cubits and fingers were used during the periods;
2. Different values of the zero point were used during the period;
3. Differences between the solar and lunar calendars.

In the original compilation, Toussun was of the opinion that all data must correspond to the scale engraved on the nilometer, where all cubits were divided into 24 fingers. However, al-Hijazi and Ibn Taghri Birdi gave 62 cases of the minima in which the number of fingers was between 24 and 27. Toussun disregarded those data as erroneous and changed them to a

24-finger scale. Later, Popper attempted to compile and correct the errors based on the conclusion that the scale of the nilometers has varied and between AD 641 and 1522 (which is the period examined and documented by both al-Hijazi and Taghri Birdi) two scales were in use:

1. Between 1 and 12 cubits, one cubit had 28 fingers and one finger was equal to 0.0192 m;
2. Between 13 and 21 cubits, one cubit had 24 fingers and one finger was also equal to 0.0192 m.

In addition, three different scales were used on the nilometer for the periods AD 641–1522, AD 1523–1860 and after 1861. Apart from that, each scale had a different zero point (8.15, 9.77 and 8.81 m, respectively).

Besides the difference in scale, compilation of the records is complicated by leap years. Arab historians used to date all events according to the Muslim lunar calendar, which contains 354 days (Brockelmann, 1953) while fiscal matters were based on the solar calendar. Thus in 33 years, the difference between the lunar and solar calendar amounted to 1 'leap year' (Wustenfeld, 1854). Correction of the calendar by leaving out one year after 33 years was an important decision, and only the Caliph and, later, the Sultan were entitled to change it.

The impact of leap years on the records was considered in different ways. Sometimes the historians adhered to the Caliph's edicts, at others the Muslim year was left in the records, and thus Taghri Birdi provided 33 sets of data for 32 solar years. An attempt to correct the data by eliminating duplicated values was made by Popper.

The records may be also influenced by a steady rise in the Nile riverbed which, according to various authors (see Wilcocks, 1889), varied between 0.132 and 0.234 m. No definite conclusion, however, can be made as to whether this phenomenon is a continuing trend or a periodical feature. Errors based on changing units, difference between the Muslim and Christian calendars and the steady rise of the Nile riverbed had to be eliminated.

No doubt the records are based on the observations of the famous nilometer situated at Rauda (Borchardt, 1906). There have been disagreements among Egyptologists on the question as to whether the nilometer (which still exists) comes from ancient Egypt or whether it was constructed by Arabs. Egyptologists have proved the existence of at least 31 nilometers along the river. However, Sethe (1905) considered Rauda as a sanctuary of the 'House of Hapy', identified by Egyptians as the God of the Nile. According to Popper, a systematically written record had to exist, because at the time of flood culmination the river level was announced in the streets of Cairo together with that of the preceding year, probably as a basis for tax regulation. However, the original records probably were presented in a different manner.

## 9.4 Analysis of the records

The records available for the analysis are shown in Table 9.4. A series of statistical tests was applied to the above sequences in an attempt to separate periodical and random components. The method of hidden periodicities (Balek, 1989) is based on the presumption that the occurrence of more periodicities existing in a sequence is not detectable by standard tests at the same level, as is the case in which the sequence contains one periodical component. Anderson's test (1942) was employed to determine the significance of the correlation coefficient, and determines whether the sequence is formed by independent random variables with the same distribution.

**Table 9.4**  Analysed records (commencing AD 646)

| Source | Taghri Birdi | | Al Hijazi | |
|---|---|---|---|---|
| Sequence Unit | minima m | maxima m | minima m | maxima m |
| Length | 663 | 849 | 849 | 849 |

A Fisher test (Hannah, 1964) examined the significance of periodical components in the sequence. Spectral density was calculated by Parzen's formula (Granger and Hatanaka, 1964) and finally the order of autoregression was determined by Whittle's test, which can trace the autoregression of higher degrees. By using the above tests, the hidden periodicities shown in Table 9.5 have been traced in the Nile minima and maxima sequences (Andel *et al.*, 1971).

In modern records compiled by Hurst (1954) periods of 84, 22 and 7 years were found significant. Obviously, the longer periods in the sequences may indicate some trends as, for example, produced by the shifting of the riverbed. Longer periods can be considered disputable, but similar long periodical components have been found in all sequences except Toussun's maxima.

Lags significant for the autoregressive models at the 90% level of significance are shown in Table 9.6. These lags are coincidental with short-term periodical components.

An interesting result has been obtained from a comparison of the graph of the records of Nile minima and corresponding fluctuation of solar activity compiled by Link (1964), who analysed earth phenomena presumably influenced by sun radiation (Figure 9.1). The coincidence of two independently compiled sequences is remarkable.

Controversial opinions can be found in the literature on the impact of solar activity on river regimes. The formation of a river regime is not

**Table 9.5**  Periodical components (years)

| Source | Taghri Birdi | | | | Toussun | | | |
|---|---|---|---|---|---|---|---|---|
| Value | Min. | Range | Max. | Range | Min. | Range | Max. | Range |
| Length (years) | 663 | | 849 | | 849 | | 849 | |
| Periods and ranges | | 440 | | 556 | | 556 | | 556 |
| | 221 | 189–265 | 282 | 242–339 | 282 | 242–339 | 282 | 242–389 |
| | 6.6 | | 141 | 132–151 | 106 | 89–113 | 141 | 132–151 |
| | | | | | 14 | | 77 | 74–82 |
| | | | | | | | 18 | 16–18 |

**Table 9.6**  Significant lags in autoregressive models

| Source | Taghri Birdi | | Toussun | |
|---|---|---|---|---|
| Value | Max. | Min. | Max. | Min. |
| Lags | 1 | 1 | 1 | 1 |
| | 2 | 2 | 2 | 2 |
| | 3 | 3 | 3 | 3 |
| | 4 | 4 | 4 | 4 |
| | 6 | 5 | 5 | 5 |
| | 7 | 6 | 6 | 6 |
| | 8 | | 7 | 7 |
| | 9 | | 8 | |
| | 10 | | 9 | |
| | | 11 | | |
| | 13 | 13 | 15 | |
| | 17 | | | |

affected by solar activity directly but through other agents such as hydrospheres of that basin troposphere and perhaps even higher layers not only above the basin but, in most cases, in distant areas. Direct impact of solar activity can be decreased by local influences, particularly in small basins, and these can be perhaps more easily regulated through controlling man's activities. Thus various combinations of mutually interacting factors can decrease or increase extraterrestrial impacts and also develop chronological effects.

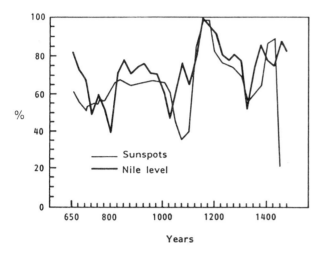

**Figure 9.1**   *Fluctuations of sunspots and Nile level* AD *663–1440*

For the sequence of Nile minima an autoregressive model has been developed as:

$$y_t = 0.42y_{t-1} + 0.1y_{t-2} + 0.06y_{t-3} + 000.01y_{t-5} + 0.02y_{t-6}$$
$$+ 0.4y_{t-11} + 0.11y_{t-13}$$

where $y = (x - 10.80)$ m and $x$ is the observed value in a particular year. Similarly, for maxima:

$$y_t = 0.2y_{t-1} + 0.07y_{t-2} + 0.09y_{t-3} + 0.06y_{t-4}$$
$$+ 0.04y_{t-6}$$
$$+ 0.02y_{t-7} + 0.05y_{t-8} + 0.08y_{t-9} + 0.06y_{t-10} + 0.1y_{t-13}$$
$$+ 0.09y_{t-17}$$

where $y = (x - 16.99)$ m.
Figure 9.2 shows the fluctuation of the river level by using the above facts and observations. A dimensionless scale provides a general picture between the period of the Old Kingdom (3240 BC) and AD 1500. Verner (1972) expected that under specific conditions such as tracts of periodically

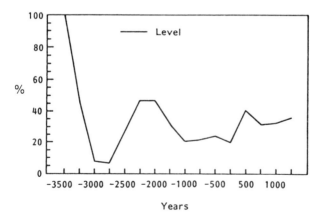

**Figure 9.2**   *Reconstruction of the Nile level, 3500* BC–AD *1440*

inundated land on the Nile banks and climatic stabilization the Nile fluctuated more or less periodically, and became the fundamental regulator of economic life. He concluded that positive regulation must have existed between the periods of an increased water level and cultural development and vice versa.

## 9.5 Climate changes traced through the water balance of the lakes

In Algeria and on the rim of northern Sahara there are numerous lakes (called *chotts*) in an advanced state of deterioration. The present drainage network of ephemeral to intermittent streams (*oueds*) indicates that much more humid conditions than today have existed in the region. The former lake levels of the Tunisian Chott el Jerid and the location of the remnants of Roman ruins in the area was discussed by Hollis in 1983. Relatively less well known are the regimes of chotts in the Algerian interior. Systematic observation of the chotts is not available, although they are a vital source of water to local agriculture.

Some of the chotts are still part of small-scale irrigation schemes. Naturally, the question must be raised as to whether the lake regimes can be improved so that more water becomes available.

One of the areas suitable for potential development is an alluvial pan south-west of Constantine with a system of chotts (Guerrah Ank Djemel and Guerah el Guelit). Both lakes were probably once part of a large interior lake between Ain Beida and Batna. One of the chotts (known as Boulhilet) is supplied by an oued bearing the same name. The oued is a part of rather complicated and bifurcating system source at Chemora (Balek, 1988).

The present drainage area for the Boulhilet was determined as 28 km². The chott is located at the rather high altitude of 832 m. Nevertheless, annual precipitation in the basin is at present only 340 mm. Monthly distribution of the rainfall is rather irregular, with a mean maximum daily precipitation of 36.4 mm and a mean maximum daily one (probability of 1%) of 75.8 mm. Monthly rainfall distribution as a percentage of the annual rainfall is shown in Table 9.7. The monthly distribution of runoff is based on observations at the gauging station at Chemora (Table 9.8). The runoff distribution year on year is highly variable because most of the floods occur in a single event. This also reflects the probability of occurrence of the rainfall volume based on Gumbel and Pearson distributions (Table 9.9). Distribution of monthly evaporation is shown in Table 9.10.

**Table 9.7** Monthly distribution of rainfall

| Month | J | F | M | A | M | J | J | A | S | O | N | D |
|---|---|---|---|---|---|---|---|---|---|---|---|---|
| % | 15 | 11 | 40 | 9 | 7 | 4 | 1 | 2 | 5 | 9 | 12 | 15 |
| mm | 51 | 37 | 34 | 31 | 24 | 14 | 3 | 7 | 17 | 31 | 41 | 50 |

**Table 9.8** Monthly distribution of runoff

| Month | J | F | M | A | M | J | J | A | S | O | N | D |
|---|---|---|---|---|---|---|---|---|---|---|---|---|
| % | 10.2 | 8.8 | 11.3 | 17.7 | 18.1 | 6.4 | 2.3 | 2.7 | 4.2 | 4.2 | 6.8 | 7.3 |
| milliard m³ | 0.046 | 0.04 | 0.06 | 0.07 | 0.08 | 0.03 | 0.1 | 0.01 | 0.01 | 0.1 | 0.03 | 0.03 |

**Table 9.9** Probability of occurrence of annual runoff

| Frequency % | 1 | 1 | 5 | 10 | 20 | 50 | 80 | 90 | 95 | 99 |
|---|---|---|---|---|---|---|---|---|---|---|
| milliard m³ | 2.44 | 1.68 | 1.2 | 0.94 | 0.70 | 0.40 | 0.15 | 0.09 | 0.04 | 0 | 0 |

**Table 9.10**   Monthly evaporation

| % | 3.3 | 4.3 | 5.8 | 6.9 | 8.8 | 12.4 | 16.8 | 15.4 | 11.7 | 6.6 | 4.4 | 3.3 |
|---|-----|-----|-----|-----|-----|------|------|------|------|-----|-----|-----|
| mm | 45 | 60 | 80 | 96 | 190 | 170 | 230 | 210 | 160 | 90 | 60 | 50 |

A water balance model (BALA 10) was developed with the aim of studying the water balance of natural and artificial reservoirs in the area. Based on monthly water balance values, among other features, it models a very important factor – morphology of the reservoirs.

Figure 9.3 shows two distinctive curves representing the relationship between the volume and surface of the reservoir. While curve A is more typical for the construction of an economical artificial reservoir, for the adaptation of a natural lake to a given environment, curve B indicates a more effective regime. In the first case a rather low increase in the water level can accommodate a large amount of floodwater while in the second the natural function of the reservoir is quite different. When inflow into the lake is high the surface area remains relatively stable. When the inflow ceases, the surface is reduced and adapts rapidly to changing conditions, and evaporation from a small surface is balanced by a low inflow.

Among other phenomena, the lake's siltation should be taken into account. Amounts of suspended solids in the tributaries are very high in

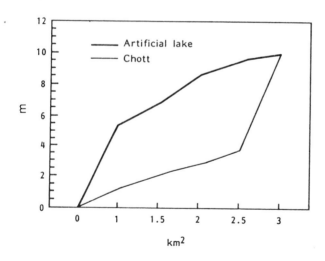

**Figure 9.3**   *Relationship surface-volume for an artificial lake and chott*

that region and thus the life span of artificial reservoirs is considered to be 30 years. Natural lakes and chotts suffer from siltation to a lesser extent.

For artificial lakes, the model (a printout of which is available from the author upon request) provides information on monthly or annual balance and determines critical months when the reservoir empties and the volume of water available throughout the balancing period is compared with actual demand. For natural lakes and chotts, the model simulated various scenarios of hypothetical climatic situations under which the lake would operate to its maximum morphological capacity. For example, for Lake Boulhilet a situation was simulated which probably existed in Roman times, and was characterized by a mean annual rainfall of 1250 mm and a mean annual temperature of 18.5°C (at present, 14.4°C).

Two water balance equations were simulated, which can be written in a simplified form:

$$340P = 16.4O + 323.6E \text{ (mm/yr)}$$

for a basin size of 28 km$^2$. For the Roman period it can be

$$1250P = 486O + 764E \text{ (mm/yr)}$$

for a basin size of 520 km$^2$ (corresponding to the former lake).

For determination of rainfall–runoff relationships five independent formulas were applied, namely, Mallet–Gauthier, Turc, Samie, the so-called 'Algerian formula' and INRH, all regionally verified. The simulated regime is very similar to the present water balance of Lake Kyoga in Uganda (Balek, 1977) – it is questionable as to the extent the basin's ecology could be found similar. It may be premature to draw conclusions based on analysis of the regime of any single lake, considering that lake shorelines in Roman ruins and the topography of the region are the only reliable data.

Nevertheless, other chotts should be examined by using the water balance simulation approach. Attention should be focused on regions with rock paintings indicating lush grasslands and hunting scenes of game now extinct – in other words, scenarios found much farther south at present. Also, areas with dense drainage networks need to be examined; they were obviously developed before the present stage of semi-aridity. From a technical point of view, it is clear that an increased volume of water available for irrigation and water supply can be obtained at present only by reducing the surface of the reservoirs.

In a doctoral thesis at London University, McClure (1984) analysed the regime of the lakes in Rub'al-Khali, which were originally in a presently arid area. He concluded that they were formed twice in recent geological history – 32 000–12 000 and 10 000–500 000 years ago. The formation of the lakes was ascribed to isolated cataclysmic rainfall. Because their depth was between 2 and 10 m and fossilized teeth of hippopotamus and bones of water buffalo, gazelle and hartbeest were found in the region, it is clear that neither the lakes nor the fauna could survive without a regular rainfall regime.

One problem remains to be solved: why has the rainfall pattern changed so radically several times in a relatively short period?

## 9.6 Conclusion

Throughout history, people have been compelled to settle in regions where water was deficient in quantity or erratic in supply. Man's endeavours to achieve a better relationship with the waters of the earth have helped to form his character. Over time, it was Nature not man which played a decisive role in climatic changes and impacts on the development of water resources. Therefore in future a difference between the impact of man and more or less natural impacts must be clearly identified.

As has been shown above, life in northern Africa has always been dependent on the seemingly uncontrollable whims of the Nile Basin and the wadis. A sufficient quantity of water has meant rich crops while water shortage has resulted in famine. The attempt to unify the disorganized system of water regulation and to build irrigation tracts was a very important (perhaps decisive) factor in the emergence and growth of state power. Improvements in the climate and a corresponding favourable hydrological regime undoubtedly played a significant role at that stage. A similar study of the fluctuation of different water regime indices with economic political and cultural development elsewhere may contribute to a better understanding of climatic fluctuation and water resources.

## Note

The designation employed and the presentation of material throughout this chapter do not imply the expression of any opinion on the part of UNEP.

## References

Andel, J. and Balek, J. (1971) 'Analysis of periodicity in hydrological sequences', *Journal of Hydrology*, **14**, 66–72.
Andel, J., Balek, J. and Verner, M. (1971) 'An analysis of historical sequences of the Nile maxima and minima', *ECA/WMO Conf. on the Role of Hydrology and Hydrometeorology in the Economic Development of Africa*.
Anderson, R. L. (1942) 'Distribution of the serial Correlation Coefficient', *Annual of Math. Stat.*, **XIII**.

Balek, J. (1977) *Hydrology and Water Resources in Tropical Africa*, Elsevier, Amsterdam.

Balek, J. (1988) 'Le régime hydrologique du Bulhilet', *Interim Report to Hydroproject est*, Constantine, Algeria.

Balek, J. (1989) *Groundwater Resources Assessment*, Elsevier, Amsterdam.

Borchard, L. (1906) *Nilmesser und Nilstandmarken*, Berlin.

Brockelmann, C. (1953) *Arabische Grammatik*, Leipzig, 208–209.

Butzer, K. W. (1959) 'Some recent geological deposits in the Egyptian Nile Valley', *Geog. Journal*, **125**, 75–79.

Chaline, J. (1985) *Histoire de l'homme et des climat au Quaternaire*, Doin, pp. 144–149.

Dixey, F. (1962) 'Geology, geomorphology and groundwater hydrology', *UNESCO Symposium on the Problems of Arid Zones*, Paris.

Fairbridge, R. W. (1962) 'Radiocarbon dates from the Nile', (cf. B. Trigger, 1965, p. 29, note 1).

Fairbridge, R. W. (1963) 'Nile sedimentation above Wadi Halfa, during the last 20 000 years', *KUSH*, **XI**, 96–107.

Gauthier, H. (1914) 'Quatre fragments nouveaux de la pierre de Palerme au Musée du Caire', *CRAIBL*, 489–496.

Geyh, M. A. and Backhaus, G. (1978) 'Hydrodynamic aspects of carbon-14 groundwater dating', *Isotope Hydrology*, IAEA, Vienna, pp. 631–643.

Granger and Hatanaka (1964) *Spectral Analysis of Economic Time Series*, Princeton University Press, Princeton, NJ.

Hannan, E. J. (1964) *Time Series Analysis*. Chapter 4 (Russian edition), State Publishing House.

Heinzelin, J. de (1967) Pleistocene sediments and events in Sudanese Nubia', in *Background to Evolution in Africa*, Chicago, Chicago University Press, pp. 313–328.

Hollis, E. (1983) Personal communication.

Hurst, H. E. (1954) *Le Nil*, Payot, Paris.

Jequier, G. (1906) 'Les nilometres sous l'ancienne empire', *BIFAO*, **V**, 63–64.

Labeyrie, J. (19??) *L'homme et le climat*, Denoel, Paris.

Langles, L. (1810) 'Table chronologique des crues du Nil les plus remarkables depuis l'an 23 de l'hegire'. *Notices et extraites des manuscrits de la Bibliothèque imperiale et autres bibliothèques publiés par L'instituteee*.

Leakey, L. S. B. (1964) 'Prehistoric man in the tropical environment', *Symposium on the Ecology of Man in the Tropical Environment*, Nairobi, Morges, pp. 24–29.

Link, F. and Linkova, Z. (1959) 'Méthodes astronomiques dans le climatologie historique', *Studia Geophysica and Heodetice*, Prague, **33**, 43–61.

McClure, H. (1984) 'Lakes of Rub'al-Khali, doctoral thesis, University of London.

Nilsson, E. (1932) 'Quaternary glaciations and pluvial lakes in British East Africa', *Geogr Ann.*, Stockholm, **13**, 249–349.

Jequier, G. (1906) 'Les nilometres sous l'ancienne empire', *BIFAO*, **V**, 63–64.

Reisner, G. A., Dunham, D. and Janssen, J. M. (1960) *Second Cataract Forts*, Vol. I, Boston University Press, Boston.

Sethe, K. (1905) *Beitrage zur altesten Geschichte Agyptens*, Leipzig.

Sethe, K. (1933) *Urkunden des Alten Reiches*, Leipzig.

Tousson, O. (1925) *Mèmoire sur l'histoire du Nil*, t. II, *Mèmoires presentés à la Société Archologique d'Alexandrie*, t. IV, Cairo.

Trigger, B. G. (1965) *History and Settlement in Lower Nubia*, Yale University Publications in Anthropology, No. 69.

Verner, M. (1972) 'Periodical water-volume fluctuations of the Nile'. *Archive oriental*, **40**, 105–124.

Vingard, E. (1923) 'Une nouvelle industrie lithique le "Sebilien"', *BIFAO*, **XXII**, 1–77.

Wayland, E. J. (1934) 'Rifts, rivers and early man in Uganda', *Royal Ant. Inst. Journal*, **64**, 332–352.

Wilcocks, W. (1989) *Egyptian Irrigation*, London.

Wustenfeld, F. (1854) *Vergleichungen Tabellen der Muhamedanischen und Christlicher Zeitrechnun*, Leipzig.

# 10  THE POTENTIAL EFFECTS OF CLIMATIC CHANGE ON WATER RESOURCES MANAGEMENT IN THE UK

Nigel Arnell
*Institute of Hydrology, Wallingford, UK*

## 10.1 Introduction

The designers and managers of water management schemes are well accustomed to dealing with seasonal and annual variations in climatic and hydrological behaviour between years, and indeed many schemes are explicitly designed to overcome problems caused by these variations. Designs and operating rules are generally based on past experience, and in particular on the experience of recent years.

Current indications, however, are that the basis for water management may be very different by the middle of the next century. Not only will there have been major changes in economic development and population levels, but human-induced changes in atmospheric composition may have stimulated very significant fluctuations in climate and hence hydrological characteristics (see Bolin *et al.*, 1987, for a comprehensive review). Long-term water resources planning requires an assessment of the possible consequences of increasing concentrations of greenhouse gases (principally carbon dioxide, methane, nitrous oxide and CFCs), and these assessments also serve to draw attention to the problems which may arise if greenhouse gases are allowed to increase without restraint.

This chapter describes some continuing investigations into the possible changes in water resource characteristics in the UK under a changed, warmer, climate (Beran and Arnell, 1989). An initial qualitative assessment has been made of potential changes to a range of water management issues relevant in the UK, and more quantitative studies are underway into water supply and resource reliability. The chapter begins with a summary describing water management issues in the UK at present and the possible types and degrees of potential changes in the climate.

## 10.2 Water resources in the UK and possible future changes

The British Isles are in the humid temperate region, and, as such, have a climate characterized by moderation and generally abundant rainfall. In common with much of northern and western Europe, there is a relatively small variation in hydrological behaviour over time compared with more arid regions, but in particular places and at certain times there may be very strong pressures on water resources.

Average annual rainfall in the UK ranges from around 550 mm in eastern England to over 2500 mm in western Scotland, and runoff varies from under 100 mm to over 1500 mm. The seasonal variation in rainfall is limited over much of the UK, but higher summer evaporation means that flows are at their minimum in late summer and early autumn. Catchment geology exerts a major control on the flow regime, and the variation in flow during a year in a chalk catchment, for example, will be very much less than that in a nearby clay catchment. Groundwater is found in virtually all the sedimentary deposits in the UK, but is most widespread in the south and east. Recharge occurs during winter, once soil moisture deficits are filled.

Despite the apparent abundance of water, however, there are several areas of concern for water managers. The south-east of England, for example, has the lowest rainfall and the driest summers but a high and increasing demand for water. The dry summer and autumn of 1989 led to supply problems in many parts of England and Wales, due not only to the dry weather but also to a long-term trend of increasing demands. Water quality problems also have high public and political profiles at present, with increases in nitrate concentrations in both surface and groundwater in some areas causing particular concern.

The current state of climate modelling is such that it is impossible to make definitive predictions of the future climate for a region as small as the UK (or, indeed, western Europe). It is possible to glean some indications from general circulation models (GCMs), but it is at present most appropriate to explore climate change impacts through future climate *scenarios*. These scenarios are not forecasts, but must represent feasible future conditions. A review of GCM estimates of the UK climate following an effective doubling of atmospheric carbon dioxide concentrations (Hulme and Jones, 1988) indicates that temperatures may rise by between 3–5°C, with a slightly greater increase in winter, but that possible changes in precipitation are less clear. Most models suggest an increase in total annual precipitation – to varying degrees – and imply that, in southern England at least, summers may become drier. In view of these uncertainties, climate change impact studies must consider several scenarios.

## 10.3 Impacts of climatic change on water management: an overview

There has been a large increase in the number of papers considering the hydrological impact of climate change since Beran's review in 1986. Some of these examine indices relevant to water resources analysis, such as flow duration curves or storage–yield diagrams (e.g. Nemec and Schaake, 1982), but there have been few attempts to determine the sensitivities of water management activities to climatic change. Notable exceptions include Schwarz (1977), Callaway and Currie (1985), and Riebsame (1988).

Table 10.1 shows a preliminary attempt to indicate the sensitivities of various water management activities in the UK to climatic change. The columns in the table summarize activities undertaken by a variety of public and private agencies, and the columns represent different dimensions of climatic change: the first three refer to changes in flow regimes. The table indicates, for example, that flood-management activities will be insensitive to changes in total annual runoff, but will be much more sensitive to changes in winter flows and soil moisture storage and more particularly extreme rainfalls.

Many of the sensitivities shown in Table 10.1 are based on 'educated guesswork', which may change when more information becomes available. The most important activities concern water supply (from both surface and groundwater sources), flood management (including reservoir safety), the management of low-lying coastal areas and the maintenance of water quality, although sensitivities depend on the type of management scheme in place (such as the site of storage, for example). Possible effects of climatic change on all these areas are discussed in Beran and Arnell (1989) and Jenkins and Whitehead (1989), but most quantitative work has been done on river flow regimes and the reliability of surface-water supply.

## 10.4 The impacts of climatic change on flow regimes and water supply

It is important to recognize at the outset that the evolution of water resource availability and scheme reliability will depend not only on climatic changes but also on economic developments, particularly with respect to demand for water. A proportion of these developments may be related to climate – perhaps demand will increase in a warmer world – and will be influenced by the reaction of policy makers to changes in resource reliability – perhaps an increasingly scarce resource will encourage more efficient water use. Land-use changes, which may be related to climatic changes (directly or via policy responses to change), may also contribute to changes in hydrological parameters and flow regimes. Figure 10.1 sum-

110

**Table 10.1** An initial assessment of the sensitivities of water management issues in the UK to climatic change

| Water management activity | Dimension of climate change 'Hydrological' | | | | |
|---|---|---|---|---|---|
| | Total runoff | Seasonal distribution | Extreme behaviour | Increased temperatures | Higher sea level |
| Surface water supplies ('small storage') | * | ** | ** | ** | – |
| Surface water supplies ('large storage') | ** | * | ** | * | – |
| Groundwater supplies | ** | ** | ** | * | * |
| Agricultural water use | ** | ** | ** | ** | – |
| Flood management | – | * | ** | – | ** |
| Urban stormwater drainage | – | – | ** | – | * |
| Reservoir safety | – | * | ** | – | – |
| Drainage of low-lying lands | ** | ** | ** | – | ** |
| Power generation | ** | ** | * | – | – |
| Water quality | * | ** | ** | ** | – |
| Effluent dilution | * | ** | ** | * | – |
| Inland navigation | * | ** | – | – | – |
| Fisheries | – | ** | * | ** | – |
| Recreation | – | * | – | * | – |
| Saline intrusion into estuaries | * | * | – | – | ** |
| Aquatic ecosystem management | – | ** | ** | ** | – |

– Little sensitivity.
* Some sensitivity.
** Very sensitive.

marizes the interrelationships between climate change and water supply, demand and resource availability.

Studies conducted so far have concentrated on the effects of climatic change on river flow regimes, and in particular on average annual runoff, monthly flow regimes and monthly flow duration curves. Three basic approaches to impact assessment are being followed (Arnell and Beran, 1989); (1) the use of 'temporal analogues'. (2) the use of regional databases containing data from many catchments under a range of catchment and climatic conditions, and (3) the application of hydrological rainfall–runoff models with climatic inputs defined by climatic change scenarios. Most attention is being paid to the last two approaches.

The effects of changes in precipitation and evapotranspiration on average annual runoff have been examined using a regional regression relationship derived from 214 small catchments in Britain (all with areas less than 500 km$^2$) during the FREND project (Gustard *et al.*, 1989). The model has the form:

$$RUNOFF = 0.97RAIN - 0.55PE - 147$$

where all the variables, representing average annual runoff, rainfall and potential evapotranspiration, are in millimetres. Several important limita-

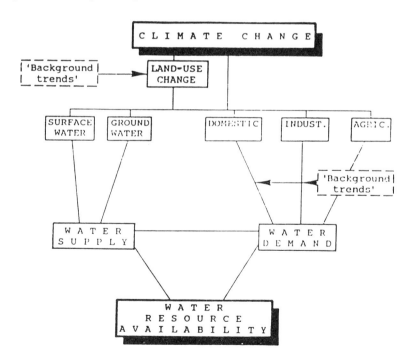

**Figure 10.1** *The interrelationships between climatic change, water supply, demand and resource availability. 'Background trends' represent changes due to evolving economic policies*

tions must be remembered. First, the model can strictly only be applied within the range of average annual rainfall and potential evapotranspiration values used in its derivation. Second, the coefficients reflect current hydrological and climatic conditions, particularly basin vegetation, which could change considerably. Finally, the effect of a given annual change in precipitation and evapotranspiration will depend on how that change is distributed through the year. The effect on runoff could be expected to be greatest if an increase in precipitation is concentrated in winter (at least in Britain). Nevertheless, a simple regional regression can provide an indication of the sensitivity of average annual runoff in a catchment to changes in rainfall and evapotranspiration.

The British model shows that, in accordance with other studies (e.g. Schaake and Chunzhen, 1989), average annual runoff is most sensitive to changes in rainfall and that a given change in rainfall triggers a greater change in runoff in dry areas (Arnell and Reynard, 1989). For example, a 10% change in rainfall in the 'dry' south-east (where the annual runoff coefficient can be as low as 20%) would give a change in runoff of approximately 30%: in the wetter north, the change would be nearer 10%. The higher sensitivity to rainfall changes in the dry south-east is important, because it is in this region that pressures on resources are greatest, and where assessments of the impacts of change are most important and also most difficult.

Changes in the characteristics of monthly flow are being assessed using several simple monthly rainfall–runoff models, including an empirical regression-based model developed by Wright (1978) and Thornthwaite-type water balance models as used by, among others, Gleick (1987). In each case, the model is being applied to several catchments covering a range of conditions in the UK. Each model is first run using the time series of recorded monthly rainfalls and average monthly evapotranspiration totals (which are assumed constant from year to year: either actual or potential evapotranspiration are used, depending on the model), and for subsequent runs the input time series are perturbed according to a defined climate-change scenario.

Figure 10.2 shows the mean monthly flow regimes as estimated using Wright's (1978) model for two lowland catchments, typical of 'responsive' and groundwater catchments, and for two climate change scenarios (Arnell and Reynard, 1989). Scenario 1 assumes a 20% increase in rainfall throughout the year: scenario 2 assumes that while rainfall is up to 20% higher in winter, spring and autumn it decreases by up to 15% in summer. Both scenarios assume the same increase in potential evapotranspiration, which is of the order of 15% higher over the year. The figure shows an increasing seasonality of flow in both catchments – reflecting the greater effectiveness of the winter precipitation increase – and the results suggest that higher winter rainfall in groundwater-fed catchments will give higher summer flows even when summer rainfall is reduced (a similar result was found by Bultot *et al.*, 1988, in Belgium). Quite similar changes in mean monthly regimes were found with different rainfall–runoff models, but the various

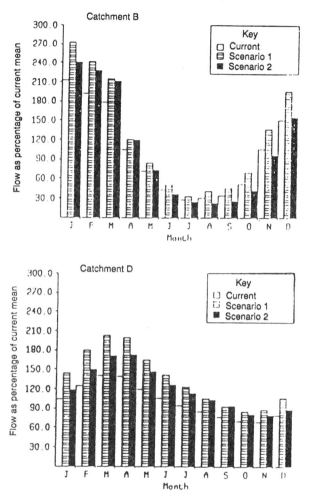

**Figure 10.2** *Mean monthly flows under current climate and two climate change scenarios. Scenario 2 incorporates a drier summer (from Arnell and Reynard, 1989). Catchment D is groundwater-fed; catchment B is more responsive*

models tended to produce different flow-duration curves. The flow-duration curves under a changed climate tended to be steeper with all models (reflecting the increased seasonality in flows), but the degree of convergence at the bottom (low flow) end and the relative change in the frequency of low monthly flows were very different in some cases. This was attributed to model formulation and parameterization, and is being further tested. The implications of runoff regime changes on the reliability of supply will depend on scheme size, and further work is necessary. Small 'one-season' reservoirs, for example, are likely to be more affected by dry summers than larger reservoirs, which have longer critical periods.

Although of a preliminary nature, the results of the studies so far have suggested two conclusions. First, the sensitivity of surface-water resources to climate change is greatest in the drier south-east of England, where pressures on resources are greatest and growing most rapidly. Second, the geological characteristics of a catchment appear to have a very important influence on the sensitivity of river flow regimes to climatic change. To a certain extent, the geological characteristics of many catchments in southern Britain may allow wetter winters to compensate for the drier summers that may occur in a warmer Britain, and maintain river flows throughout summer.

## 10.5 Future research directions

This chapter has given an overview of the potential effects of climatic change on water management in the UK, focusing in particular on water supply. Rather than summarize the conclusions further, it is most appropriate to end by indicating where further effort needs to be placed, both in the UK and elsewhere.

First, it is important that climate models are improved in order to help with the derivation of realistic scenarios for future climates. In particular, information is needed on the spatial characteristics of climatic changes and potential changes in short-term (such as daily) climatic characteristics. Spatial aspects are important because the responses of water managers to change will depend on how that change is felt across a region: the feasibility or desirability of inter-regional water transfers, for example, will depend on the relative impact of change in each region. Information on fluctuations in short time-scale climatic behaviour is necessary before any assessments can be made of possible changes in flood risk and reservoir safety.

Considerable progress has been made in estimating the impacts of change on indices of hydrological behaviour relevant to water resources analysis such as the flow duration curve and storage–yield diagram. The impact of change on water as perceived and exploited by water users, however, will depend on the nature of the water management infrastructure, such as reservoir size and operating rules. Studies of possible changes in the operation of typical water management schemes are therefore needed. While these studies would be limited at present by the nature of the assumed changes in climatic inputs, they would help to clarify, for example, the importance of reservoir size in controlling the sensitivity of reliability to change and the role of operating rules in mitigating the effects of change. Would a possible reduction in summer rainfall in a region, for example, necessitate the construction of new reservoirs, or could the changes be accommodated by relatively minor alterations in operating rules? Detailed case studies based on water-management schemes could also allow changes to be shown in different ways. It may be possible to express the effect of a climate change in terms of alterations in standards of service, such as the frequency of restrictions on hosepipe use, or in total power generation potential.

The first results of studies into particular aspects of the impacts of climate change on water management the UK indicate that it may be difficult to generalize results even within a relatively small area. The changes which may take place can be very sensitive to the nature of the climatic fluctuation in some regions, and the details of the hydrological change will depend on the geological characteristics of a catchment. It is important therefore to understand the current relationships between climate and hydrological response before attempting to make inferences about any possible changes. The nature of these relationships may influence the methodology used to estimate impacts. A simple relationship between annual rainfall and runoff may not always be appropriate, and the hydrological characteristics of some regions are very dependent on the occurrence of short-time-scale events which are currently very difficult to model in a GCM. In particular, it will be much harder to estimate the impacts of change in a semi-arid area, for example, than in relatively humid Britain.

It is important to recognize that the actual impact of climatic fluctuation on water users will be determined not just by the changes in physical resources in conjunction with the existing water management infrastructure but also by the ability of water management to respond to changing circumstances. Some institutional structures may more readily respond to changes in water availability and characteristics than others, and any serious assessment of the impact of change in a region must consider the adaptability of the water management structure. Finally, water management 'futurologists' must not concentrate on climatic changes to the exclusion of all else: other changes – in land use and demand for water, for example – are taking place, which in some regions may overwhelm the effects of changes in climate.

## Acknowledgements

The studies reported in this chapter were funded by the UK Department of the Environment.

## References

Arnell, N. W. and Beran, M. A. (1989) 'Climate change scenarios for water resource impact studies', *IIASA Task Force Meeting on Development of Regional Climate Scenarios for Impact Assessment*, 20–22 February.

Arnell, N. W. and Reynard, N. S. (1989) 'Estimating the impacts of climatic change on river flows: some examples from Britain', *Conference on Climate and Water, Helsinki*, Vol. 1, 426–436, 11–15 September.

Beran, M. A. (1986) 'The water resources impact of future climate change and variability', in Titus, J. G. (ed.), *Effects of Changes in Stratospheric Ozone and Global Climate*, Vol. 1, UNEP/EPA, Washington, DC.

Beran, M. A. and Arnell, N. W. (1989) *Effect of Climatic Change on Quantitative Aspects of UK Water Resources*, Report to DoE, Institute of Hydrology, Wallingford, UK.

Bolin, B., Doos, B. R., Jager, J. and Warrick, R. A. (eds) (1987) *The Greenhouse Effect, Climatic Change and Ecosystems: A Synthesis of the Present Knowledge*, Wiley, New York.

Bultot, F., Coppens, A., Dupriez, G. L., Gellens, D. and Meulenberghs, F. (1988) 'Repercussions of a $CO_2$ doubling on the water cycle and on the water balance – a case study for Belgium', *Journal of Hydrology*, **99**, 319–347.

Callaway, J. M. and Currie, J. W. (1985) 'State-of-the-art report: water resources', in *State-of-the-Art Report on the Impact of $CO_2$-induced Climatic Change*, US Dept of Energy, Washington, DC.

Gleick, P. H. (1987) 'Regional hydrological consequences of increases in atmospheric $CO_2$ and other trace gases', *Climatic Change*, **10**, 137–161.

Gustard, A., Roald, L. A., Demuth, S., Lumadjeng, H. S. and Gross, R. (1989) *Flow Regimes from Experimental and Network Data (FREND)*, Institute of Hydrology, Wallingford, UK.

Hulme, M. and Jones, P. D. (1988) *Climatic Change Scenarios for the UK*, Climatic Research Unit report to Institute of Hydrology.

Jenkins, A. and Whitehead, P. G. (1989) *Climate Change and UK Water Quality*, Report to DoE, Institute of Hydrology, Wallingford, UK.

Nemec, J. and Schaake, J. C. (1982) 'Sensitivity of water resource systems to climate variation', *Hydrological Sciences Journal*, **27**, 327–343.

Schaake, J. C. and Chunzhen, L. (1989) 'Development and application of simple water balance models to understand the relationship between climate and water resources', *New Directions for Surface Water Modelling*, IAHS Publ. 181, 343–352.

Riebsame, W. E. (1988) 'Adjusting water resources management to climate change', *Climatic Change*, **13**, 69–97.

Schwarz, H. E. (1977) 'Climatic change and water supply: how sensitive is the Northeast?' in *Climate, Climate Change and Water Supply*, National Academy of Science, Washington, DC, 111–120.

Wright, C. E. (1978) *Synthesis of river flows from weather data*, Central Water Planning Unit, Technical Note 26, Reading, UK.

# 11 RELATION BETWEEN RAINFALL FLUCTUATIONS AND WATER MANAGEMENT IN JAPAN

Yutaka Takahasi
*Shibaura Institute of Technology, Japan*

## 11.1 Trends in Japan's annual precipitation since 1900

The average annual precipitation in Japan, estimated from about 1000 stations over the 30-year period from 1956 to 1985, is approximately 1750 mm. Very high rainfalls are recorded during the rainy season (June–July) and the typhoon season (July–October), as can be seen in Tables 11.1 and 11.2. Japan experiences four distinct seasons. The climate in the rainy season and the summer that follows is generally hot and humid, although sharp contrasts can be seen between the coasts of the Pacific and the Sea of Japan due primarily to their separation by high mountain ranges. In winter heavy snowfalls take place in the northern regions while fine weather prevails almost daily on the Pacific coast. The rain which falls in the months of June and July provides vital irrigation water to paddy fields but also causes flooding and flood damage. From July to October, typhoons originating in the south-eastern sea cause storms and further flood damage.

Figure 11.1 shows the long-term average annual precipitation estimated from 46 stations throughout Japan since 1900 and indicates clearly a decreasing tendency after 1970. Low values of annual precipitation have been recorded in 1973, 1978 and 1984.

**Table 11.1** Maximum recorded point precipitation in one hour (mm)

|  | *Precipitation* | *Date* |
|---|---|---|
| Nagayo | 187 | 23 July 1982 |
| Kobutsu | 183 | 23 July 1982 |
| Fukui | 167 | 22 March 1952 |
| Fujinomiya | 153 | 24 August 1972 |
| Ashizuri | 150 | 17 October 1954 |

**Table 11.2**   Maximum recorded point precipitation in 24
hours (mm)

|  | *Precipitation* | *Date* |
|---|---|---|
| Hisawa | 1114 | 16 September 1976 |
| Saigo | 1109 | 25 July 1957 |
| Moriyama | 1057 | 25 July 1957 |
| Tateyama | 1016 | 19 July 1944 |
| Odaigahara | 1011 | 14 September 1923 |

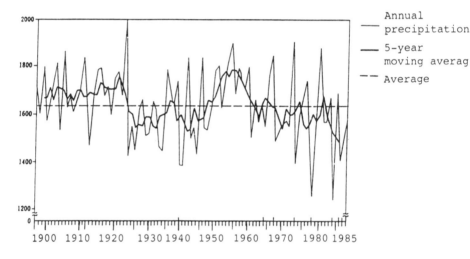

**Figure 11.1**   *Trend in long-term average annual precipitation*

The year 1984 was extremely abnormal in terms of precipitation. The lowest average annual precipitation since records began in Japan 100 years ago is observed in this year. Also, in 1984, 20 out of 80 stations in Japan recorded the lowest annual precipitation and 40 the three lowest annual values. The year 1978 was also unusual in that 27 stations observed the three lowest values.

The stations which recorded low precipitation in 1984 are concentrated in Hokkaido and Tohoku (north-eastern Honshu). As shown in Figure 11.2, annual precipitation in Hokkaido has been decreasing since 1975, except in 1981. An abnormality can also be noted in monthly precipitation since the 1970s.

Figure 11.3 shows the frequency of occurrence of abnormal monthly precipitation in each decade since the 1930s. Abnormally high monthly precipitations were recorded in the 1950s whereas low values were observed in the 1970s and 1980s. Here, abnormally high and low monthly precipitations define the maximum or the minimum in each of the past 30 years.

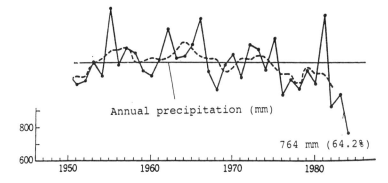

**Figure 11.2** *Trend in long-term average annual precipitation at 20 points in Hokkaido*

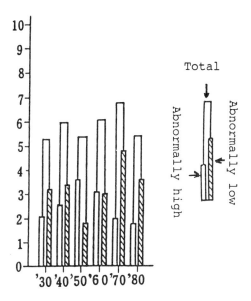

**Figure 11.3** *Frequency of abnormally high and low monthly precipitation in each decade since the 1930s*

On a regional basis, the decreasing trend in precipitation over the past 20 years can be seen in the whole of Japan, except in the Sea of Japan side of Tohoku and Hokuriku and in parts of Shikoku and Kyushu. Annual precipitation in the decade 1978–1987 was less than in the previous decade (1968–1977), as illustrated in Figure 11.4. The ratio of the standard deviation of annual precipitation in the decade 1978–1987 to that in the previous decade is shown in Figure 11.5. Fluctuations have become greater in recent years, especially in the Kanto, Tokai and Northern Kyushu areas.

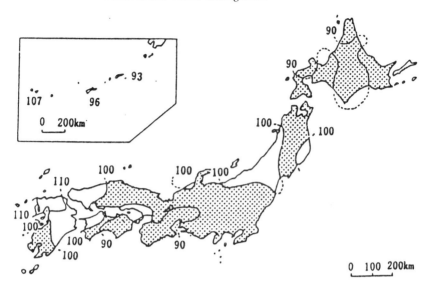

**Figure 11.4**   *The ratio of annual precipitation in the decade 1978–1987 to that in the previous one (1968–1977)*

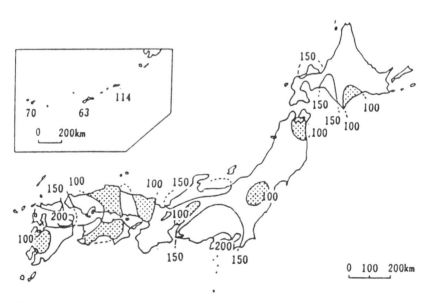

**Figure 11.5**   *The ratio of standard deviation of annual precipitation in the decade 1978–1987*

Table 11.3 gives a comparison of the abnormal climatic conditions that prevailed in Japan in the five-year period 1984–1988 with those in other parts of the world.

Figures 11.6 and 11.7 show the decreasing trend of an index of potential water resources quantity, which is defined as follows:

Potential water resources quantity index
= area (annual precipitation − annual evaporation)

The index has lower values for the recent 30 years. The drought that occurred once in 10 years in 1956–1975 took place once in four years in 1966–1985 (Figure 11.6). All regions except Okinawa have lower values of the index in recent years, as shown in Figure 11.7.

**Table 11.3**  Notable abnormal events 1984–1988

|  | Japan | World |
|---|---|---|
| 1984 | Severe cold spell in winter Severe heat wave in summer | Severe droughts in Ukraine and Africa |
| 1985 | Severe heat wave in summer | Cold summer in northern Europe and cold weather in all Europe |
| 1986 | Low precipitation in autumn in western Japan | Severe drought in south-eastern USA and low temperatures in northern Europe |
| 1987 | Warm winter, low precipitation in spring | Drought in India, severe flood damage in Bangladesh and heat wave in Greece |
| 1988 | Long rainy season | Drought in central western USA, heat wave in southern China, floods in Bangladesh |

## 11.2  Recent water shortages in Japan

Recent drought (or water shortage) events can be divided into three categories.

1. Water shortage due to a small river basin. Dams and reservoirs that regulate the availability of water cannot be constructed in small basins. When the rainfall is low such regions experience water shortages (e.g. several islands in the Seto Inland Sea, a peninsula in the Nagasaki Prefecture).

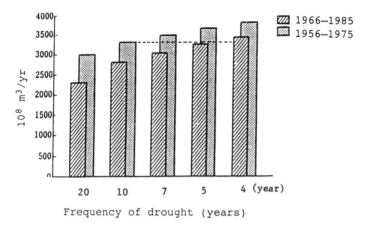

**Figure 11.6**   *Frequency of drought occurrence shown by the potential quantity of water resources in a comparison between 1966–1985 and 1956–1975*

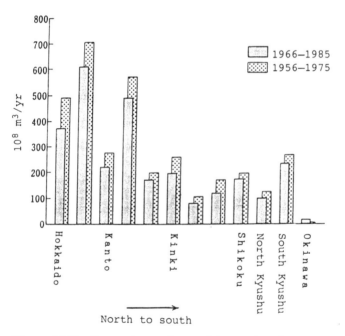

**Figure 11.7**   *Comparisons of potential quantity of water resources in each district between 1966–1985 and 1956–1975*

2. Water shortage caused by rapidly increasing demand and delayed supply. There is a time lag between demand and supply due to delayed development of water resources (e.g. the Saitama Prefecture near Tokyo in 1984, 1985 and 1987). The water-utilization rights from rivers in the newly urbanized areas near Tokyo are also uncertain because of delayed development.

3. Water shortage caused by low precipitation in the long term. In this case the actual river discharge may fall below that needed to maintain water rights asbstractions (e.g. the drought in Fukuoka in 1978).

In general, water shortages may occur as a result of the combined effect of any two of the above three causes. Recent low precipitations make the shortages even more acute. Drought events for 1964–1987 are listed in Table 11.4.

**Table 11.4**  Drought events 1964–1987

| Year | Place | Restriction period on water supply (days) | Cumulative deficit ratio (%.day) |
|---|---|---|---|
| 1964 | Tokyo municipal area | 84 | 7660 |
| 1967 | Kita-Kyushu | 130 | 1450 |
| | Tsukushino | 22 | 1450 |
| | Nagasaki | 72 | 5000 |
| 1973 | Matsue | 135 | 3930 |
| | Takamatsu | 58 | 2390 |
| | Naha | 239 | 2650 |
| 1977 | Osaka | 134 | 1500 |
| | Naha | 176 | 2401 |
| 1978 | Fukuoka | 287 | 8160 |
| | Kita-Kyushu | 173 | 2090 |
| | Osaka | 159 | 1800 |
| 1981 –82 | Naha | 326 | 6210 |
| 1984 | Gamagori | 154 | 2170 |
| | Tokai | 113 | 2280 |
| | Osaka | 156 | 2900 |
| 1986 | Gamagori | 153 | 2270 |
| | Tokai | 146 | 1200 |
| 1987 | Tokai | 180 | |
| | Saitama | 71 | |
| | Gamagori | 274 | |

The cumulative deficit ratio, an index for evaluating the severity of a drought, is defined as follows:

$$= \frac{\text{Deficit discharge}}{\text{Target discharge}} \times (\text{number of days}) \times 100\%$$

$$= \frac{\text{Target discharge} - \text{actual discharge}}{\text{Target discharge}} \times (\text{number of days}) \times 100\%$$

The discharge referred to above is for water supply and the unit of the index is the percentage-day. Summation in the above calculation is done on a daily basis during a drought. The relationship between the magnitude of the index and the severity of the drought is as follows:

1. $> 1000\% \cdot$ day – signs of damage begin to appear
2. $> 2000\% \cdot$ day – damage becomes severe in both domestic and industrial activities.

## 11.3 Periodicity in the occurrence of severe storm rainfall

Although long-term rainfall data are unavailable there appears to be a periodicity of a few decades in the occurrence of severe storm rainfalls. For example, in almost every year during the period 1945–1959 severe typhoons caused heavy rainfall leading to severe flood damage, and there were also several instances of severe flood damage between 1889 and 1910. Although information on the hydrometeorological conditions that produced such events is not available, the strong rainfall intensities of these events are beyond doubt.

Although again there are no scientific hydrological data, there is some historical evidence of severe flood damage around the year 1800 and after. For example, severe floods occurred in the Tone River Basin which runs across the Kanto Plain, including the Greater Tokyo Metropolitan Area in 1786, 1791 and 1800, as well as in 1889, 1896, 1910, 1947 and 1948. These facts suggest that there is some periodicity in the occurrence of strong typhoons with severe precipitation.

# 12 THE VULNERABILITY OF RUNOFF IN THE NILE BASIN TO CLIMATIC CHANGES

Peter H. Gleick
*Pacific Institute for Studies in Development, Environment, and Security, USA*

'The next war in our region will be over the waters of the Nile, not politics' (Egypt's Minister of Foreign Affairs, 1985).

## 12.1 Introduction

The freshwater resources of the Nile are widely shared and heavily used. Among the many benefits are irrigation, navigation, drinking water, waste disposal, natural habitats for fish, wildlife, and birds, and the generation of hydroelectricity. Growing population pressures and increasing development along the Nile are already straining the limited freshwater resources of the basin and future climatic changes will add further pressures with widespread political ramifications.

Because of the importance of the Nile for the countries of north-eastern Africa, any changes in river flow have enormous impacts. The recent decade of drought is a painful reminder of Egypt's vulnerability to changes in Nile flow. Yet the problems caused by the natural variations in Nile runoff may soon be compounded by global climatic changes that affect evaporation, precipitation and, ultimately, water availability. These climatic changes – 'the greenhouse effect' – must lead to a re-evaluation of water allocations and methods for resolving disputes over shared resources. Without such a re-evaluation, the risk of frictions and conflicts in the region will increase.

## 12.2 Future climatic changes

Global climatic changes due to a wide variety of human activities now appear inevitable. As a consequence of production of trace gases that trap heat in the atmosphere, we can expect higher global temperatures, changes in precipitation patterns, altered storm frequencies and intensities, increasing sea level and the disruption of a variety of other geophysical conditions. While debate continues about the precise details of regional climatic changes and the rate at which changes will occur, there is a growing consensus that we must begin planning today for growing climatic uncertainty.

Evidence for the nature of future climate conditions and our vulnerability to these comes from several different sources, including reconstructions of past 'paleoclimatic' conditions, model estimates of future climatic conditions and studies of climatic vulnerabilities. Unfortunately, a clear view of the future is obstructed by many uncertainties about the physical behaviour of the atmospheric system, by poorly understood interactions between the atmosphere and the oceans and by unanswerable questions about human behaviour and actions. Nevertheless, the available evidence suggests that significant and unanticipated changes are developing.

## 12.3 Climate and water resources

One of the most important consequences of future changes in climate will be alterations in regional hydrologic cycles and subsequent effects on the quantity and quality of regional water resources. Yet these consequences are poorly understood. Recent hydrological research strongly suggests that plausible climatic changes caused by increases in atmospheric trace-gas concentrations will alter the timing and magnitude of runoff and soil moisture, change lake levels and groundwater availability, and affect water quality. Such a scenario raises the possibility of considerable environmental and socio-economic dislocations, and has widespread implications for future water resources planning and mangement.

The long construction times and subsequent lifetimes of reservoirs, dams and water-transfer facilities means that planning should begin today for changes that may not become evident for years. Yet changes in water resources management and planning will only come if those responsible for our water systems can be convinced that the problem of climatic change is sufficiently real and pressing to require their attention. In international river basins, this problem is further compounded by shared responsibility and complicated (or missing) mechanisms for resolving disputes over water availability.

## 12.4 Shared international river basins

Water flows without regard for political boundaries; indeed, most freshwater resources are shared by two or more nations. The immediate result of sharing water resources may be uses of water that cause frictions with other, downstream, users; an ultimate result may be conflict and war.

Forty-seven per cent of all land area falls within international river basins and over 200 river basins are multinational, including 57 in Africa and 48 in Europe (se Tables 12.1 and 12.2). The extent of this interdependence can be seen in Table 12.3, which lists thirteen different major rivers with five or more nations forming part of the watershed. Examples from the different continents include the Danube, whose basin lies within twelve different nations; the Nile, which runs through nine; the Amazon, which runs through seven; the Mekong, which runs through six; and the Ganges-Brahmaputra, which runs through five (United Nations, 1978).

**Table 12.1**  Number of countries with greater than 75% of total area falling within international river basins

| | |
|---|---|
| Africa | 23 |
| North and Central America | 0 |
| South America | 6 |
| Asia | 8 |
| Europe | 13 |

Source: United Nations (1978).

**Table 12.2**  International river basins

| | *International river basins* | *Percentage of area in international basins* |
|---|---|---|
| Africa | 57 | 60 |
| North and Central America | 34 | 40 |
| South America | 36 | 60 |
| Asia | 40 | 65 |
| Europe | 48 | 50 |
| Totals | 215 | 47 |

Source: United Nations (1978).

Regions with a history of international tensions or competition over water resources include the Jordan and Euphrates rivers in the Middle East, the Nile, Zambezi, and Niger rivers in Africa, the Ganges in Asia and the Colorado and Rio Grande rivers in North America. As water demands rise due to growth in population and industry, the probability of conflict over remaining water resources will also increase (Gleick, 1989a).

Future climatic changes can reduce or exacerbate these water-related tensions. Among the critical concerns are changes in (1) water availability from altered precipitation patterns or higher evaporative losses due to

**Table 12.3**   Rivers with five or more nations forming part of the basin

|                    | #  | Area (km²)  |
|--------------------|----|-------------|
| Danube             | 12 | 817 000     |
| Niger              | 10 | 2 200 000   |
| Nile               | 9  | 3 030 700   |
| Zaire              | 9  | 3 720 000   |
| Rhine              | 8  | 168 757     |
| Zambezi            | 8  | 1 419 960   |
| Amazon             | 7  | 5 870 000   |
| Mekong             | 6  | 786 000     |
| Lake Chad          | 6  | 1 910 000   |
| Volta              | 6  | 379 000     |
| Ganges-Brahmaputra | 5  | 1 600 400   |
| Elbe               | 5  | 144 500     |
| La Plata           | 5  | 3 200 000   |

Source: United Nations (1978).

higher temperatures, (2) the seasonality of precipitation and runoff, (3) flooding or drought frequencies, and (4) the demand for and the supply of irrigation water for agriculture.

### 12.4.1 The Nile Basin

The Nile is an international river of considerable regional importance. It flows through some of the most arid parts of northern Africa and is vital for agricultural production in Egypt and the Sudan. As a result, the Nile is extensively used – so extensively that almost no water reaches the Mediterranean. Rational utilization of the Nile is complicated by the fact that the catchment and riparian drainage area of the river is shared by nine nations: Egypt, the Sudan, Ethiopia, Kenya, Tanzania, Zaire, Uganda, Rwanda and Burundi. Although the principal water users are Egypt and the Sudan, runoff in the river is mostly generated by precipitation in other countries, particularly Ethiopia.

Competition for the waters of the Nile arose in the early 1900s over growing Egyptian needs, and the question of allocation of the Nile flow was submitted to international mediation in 1920. In 1929, a first Nile Waters Agreement was adopted. This agreement was modified in late 1959 and a new 'Agreement for the Full Utilization of the Nile Waters' was signed. Unfortunately, the 1959 treaty provisions have a number of shortcomings that will complicate resolving climate-related problems (Goldenman, 1989):

1. Seven of the nine riparian countries are not party to the treaty;
2. Water allocations are fixed, rather than proportional;

3. The Permanent Joint Technical Committee (PJTC) set up by the treaty has little authority;
4. The treaty contains no dispute-resolution procedures and no process for amendment; and
5. The provision for drought-induced shortages is vague and difficult to implement.

These ambiguities and omissions could result in a revival of frictions if the runoff available in the Nile were to be reduced by climatic changes, or if the frequency and intensity of flooding in the basin were to increase. In fact, as discussed later, there is some evidence that the greenhouse effect could lead to both a reduction in overall streamflow in this region and an increase in the intensity of seasonal floods.

In the future, the other basin nations are likely to play a larger role in the use of the Nile because of growing populations and irrigation demands (see Table 12.4). Ethiopia, for example, has already announced that it reserves the right to develop hydrologic projects on the Blue Nile and in headwater tributaries to the Nile in the Ethiopian highlands, and it has called for new basin-wide agreements to allocate shares among the other riparian nations (Waterbury, 1979). Any climatically induced change in water availability will complicate the future use of the Nile even more, contributing to political jousting and friction. The nine basin states need to resolve the ambiguities over the allocation of shortages in the basin and to reach a specific agreement about how climatic changes are to be identified and handled before such changes appear.

## 12.5 Future hydrologic conditions in the Nile Basin

Projections of changes in water availability due to the greenhouse effect come from several sources, including reconstructions of past climatic conditions that may be similar to future conditions, studies of present climatic variability, and model estimates of future changes. Each of these techniques has advantages and limitations (see, for example, Gleick, 1989b), and none of them yet can provide accurate predictions. Nevertheless, insights into future conditions and sensitivities can be gained by considering each approach.

### 12.5.1 Paleoclimates of the Nile Basin

Reconstructions of climatic conditions in the Nile Basin over the last 25 000 years, summarized in Table 12.5, show extensive dry and wet periods compared to the present. These changes are attributed to changes in the earth's orbit, the composition of the atmosphere, and the output of energy from the sun. The Holocene – from about 8000 to 5000 years BP – has been

**Table 12.4** Demographic details: Nile Basin countries

| Country | Area (000 km²) | 1989 Population (millions) | Population growth rate (%/yr) | Irrigated area arable cropland (10⁶ ha) | Total arable cropland (10⁶ ha) |
|---|---|---|---|---|---|
| Egypt | 1000 | 51 | 2.3 | 2.5 | 2.5 |
| Sudan | 2376 | 24 | 2.9 | 1.9 | 12.5 |
| Ethiopia | 1101 | 49 | 2.8 | 0.14 | 13.9 |
| Uganda | 200 | 18 | 3.5 | 0.01 | 6.6 |
| Kenya | 569 | 24 | 4.2 | 0.05 | 2.4 |
| Tanzania | 886 | 26 | 3.7 | 0.1 | 5.2 |
| Burundi | 26 | 5 | 2.8 | 0.07 | 1.3 |
| Rwanda | 25 | 7 | 3.4 | <0.01 | 1.0 |
| Zaire | 2268 | 34 | 3.0 | <0.01 | 6.6 |

1 ha = 2.38 feddans = 2.47 acres.
1 feddan = 1.038 acres = 0.42 ha.
100 ha = 1 km².

Source: World Resources Institute (1988).

**Table 12.5**   Paleoclimatic conditions in the lower Nile Basin

---

*Egypt*
25 000 BP–18 000 BP: Dry
18 000 BP–5000 BP: Heavy winter rains; increased Blue Nile flow
10 000 BP–8000 BP: Wet
6000 BP–4500 BP: Wet
2500 BP–present: Drying

*Sudan*
20 000 BP–15 000 BP: Dry
12 000 BP–7000 BP: Wet
6000 BP–3000 BP: Fairly wet
3000 BP–present: Drying

---

Sources: Butzer (1971), Wickens (1975), Shahin (1985).

suggested as a suitable analogue for future greenhouse conditions of about a 1°C warming (Budyko and Izrael, 1987). During this period, evidence suggests that wetter conditions prevailed in the lower portion of the basin. Less information is available, however, on how the hydrologic conditions in the headwaters regions changed during this period, and there are unresolved questions about the suitability of using it as an analogue for future periods.

More recently, reconstructions of lake levels, precipitation patterns and runoff in northern Africa over the last three centuries have revealed patterns of wet and dry periods throughout the record, including periods of severe drought (Folland and Palmer, 1985). Evidence links precipitation in this region with larger-scale climatic events, including worldwide sea-surface temperatures and El Niño/Southern Oscillation (ENSO) events. The failure of the seasonal rains can lead to catastrophic consequences for agricultural productivity in Sudan and Ethiopia and a reduction in runoff in the Blue Nile due to decreases in rainfall in the Ethiopian highlands.

### 12.5.2 Recent climatic anomalies

Recent climatic extremes in the Nile Basin are also useful for providing insight into the vulnerability of water users in the basin. Twice during the last 20 years, precipitation levels dropped in the headwater regions of the Nile, causing severe drought conditions. In 1987–1988 water levels in the Nile were at their lowest in recorded history and Egypt was unable to draw its entire allotted share (Ross, 1987). It is not yet possible to say with certainty whether these anomalies are examples of short-term weather variability or the first indications of longer-term persistent changes. What is known, however, is that these droughts led to significant hardship in lower basin states. They also showed the inadequacy of plans to allocate such shortages: the Permanent Joint Technical Committee, which was given authority in the 1959 treaty to draw up plans for allocating drought shortages, failed to do so.

### 12.5.3 Climate models

Large-scale computer models of the global atmospheric system – so-called general circulation models (GCMs) – attempt to reproduce the physical behaviour of the earth–atmosphere–ocean system. Despite the limitations to these models, including poor spatial resolution and simplified hydrologic parameterizations, considerable success has been achieved in mimicking past and present climates, and such models are now used to try to understand how human activities, especially the emission of greenhouse gases, will affect future climatic behaviour. There are five principal GCMs today: the Goddard Institute model (GISS), the Geophysical Fluid Dynamics Laboratory model (GFDL), the National Center for Atmospheric Research model (NCAR), the Oregon State University model, and the UK Meteorological Office (UKMET) model.

Runoff estimates generated by these models are computed in a very simplistic manner, typically using the difference between precipitation and evaporation for any grid box. Evaluating these internally generated runoff estimates from three GCMs of the region of the Blue Nile shows mixed conclusions: runoff increases in two of the models and decreases in the third. Estimates are equally mixed for other regions of the Nile Basin. While such results can be useful for suggesting where problems may occur, the GCM runoff estimates must be used with extreme caution for any other purpose (WMO, 1987).

A more realistic approach is to use the meteorologic information generated by GCMs to drive a more detailed hydrologic model of the basin. This is discussed below.

## 12.6  A simple hydrologic assessment

The best way to obtain more detailed regional information on the implications of climatic changes for water availability is to combine meteorological information on future climatic changes with the use of regional hydrologic models (Gleick, 1989b). Many different techniques are available for this type of analysis, and each has advantages and limitations, depending on the purpose for which they were designed and are to be used. First-order water balance models, for example, can provide information on the sensitivity of runoff to changes in temperature and precipitation, but cannot give data on changes in the seasonality of runoff, snowfall and snowmelt behaviour, and groundwater and soil moisture. More complicated models can provide this information, but at the expense of computer time and the need for additional hydrologic input data.

For the purposes of this analysis, a simple annual water balance model was developed to assess the vulnerability of Nile runoff to changes in precipitation and evapotranspiration. Details on such changes were provided by general circulation model output for the region, as shown in Figure 12.1.

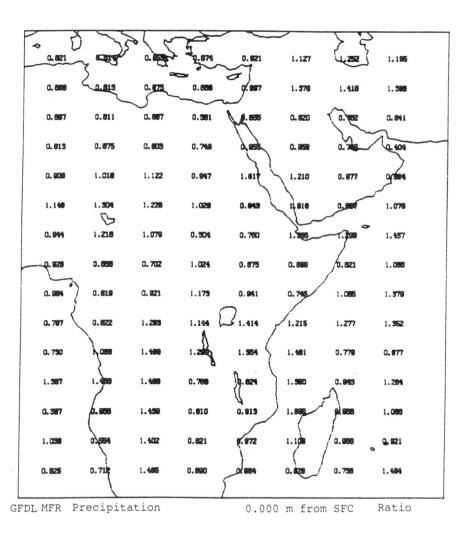

GFDL MFR  Precipitation                    0.000 m from SFC      Ratio

**Figure 12.1** *An example of a GCM's output from the Geophysical Fluid Dynamics Laboratory (Princeton, New Jersey). The figure shows the ratio for the doubled $CO_2$ climate run to the precipitation for the control run ($2 \times CO_2 \div 1 \times CO_2$) for March*

If non-evaporative losses are small, the mean annual water balance for a river can be described as:

$$R = P - E \tag{12.1}$$

where $R$ is annual runoff, $P$ is annual precipitation and $E$ is annual evapotranspiration. Under conditions of changing climate

$$P_1 = \alpha P_o \tag{12.2}$$

$$E_1 = \beta E_o \tag{12.3}$$

where $P_1$, $E_1$ are the changed precipitation and evapotranspiration at time $t = 1$, and $P_o$, $E_o$ are present precipitation and evapotranspiration. According to Wigley and Jones (1985), $\beta$ can be considered as the product of three factors, $\beta_1 \cdot \beta_2 \cdot \beta_3$, where $\beta_1$ is the change due to a change in climate, $\beta_2$ the change due to a change in vegetated area, and $\beta_3$ is the change due to the direct effects of $CO_2$ on transpiration from a vegetated surface. Another ratio, the 'runoff ratio', $\Gamma$, is defined as $R_o/P_o$. Rewriting equations (12.1)–(12.3) gives

$$\frac{R_1}{R_o} = \frac{\alpha - (1 - \Gamma_o)\beta}{\Gamma_o} \tag{12.4}$$

In arid regions, where the runoff ratio is very low, this equation is of limited use and must be employed with extreme care. Indeed, nonsensical results can be obtained for some basins for the magnitude of changes in precipitation and evaporation estimated for the greenhouse effects. For rivers such as the Nile, where much of the annual runoff is generated in highland, mountainous areas with higher runoff ratios, equation (12.4) can be used to evaluate the relative sensitivity of mean annual runoff to changes in precipitation and evaporation.

In order to do this, the Nile Basin was split into six sub-basins, shown in Table 12.6. In three of these sub-basins (the Sudd swamps, the White Nile between the Sobat River and Khartoum, and the lower Nile from Aswan to the Mediterranean Delta) the runoff ratio is very low and very little runoff is generated. Equation (12.4) is therefore not applicable to these sub-basins. Indeed, in these regions evapotranspiration considerably exceeds precipitation, and net runoff is zero or negative.

In the other three regions (the Upper White Nile, the Sobat River, and the Blue Nile and Atbara regions) substantial runoff is generated. Using the runoff ratios in Table 12.6, and estimates of changes in temperature and precipitation from one of the GCMs (the Geophysical Fluid Dynamics Laboratory model) changes in mean annual runoff were produced. In the Upper White Nile region, runoff increased by approximately 20% because of an increase in basin precipitation, while in the Blue Nile/Atbara basin, it

**Table 12.6** Nile climate change runs: simple water balance assumptions

| Nile sub-basin | Area (000 km²) | Runoff ratio[a] | Change in evapotranspiration[b] | Change in runoff $R_1/R_0$[c] |
|---|---|---|---|---|
| 1. Upper White Nile | 466 | 0.25 | 1.02 | 1.19 |
| 2. Sudd Swamps | 438 | – | 1.04 | Assumed 1.0 |
| 3. Sobat Basin | 187 | 0.25 | 1.02 | 0.95 |
| 4. White Nile to Khartoum | 343 | – | 1.00 | Assumed 1.0 |
| 5. Blue Nile/Atbara | 404 | 0.25 | 0.94 | 0.32 |
| 6. Aswan to Delta | 1042 | – | 1.04 | Assumed 1.0 |
| Total | 2880 | | | 0.67 |

[a]Estimates of the runoff ratios come from historical data and from Shahin (1985). Runoff ratios in regions 2, 4, and 6 are under 0.10 or smaller, and cannot be used in this type of analysis.

[b]Values of $\beta_1$, $\beta_2$, and $\beta_3$ varied from 0.8 to 1.15. Total evapotranspiration ($\beta$) in higher-altitude, heavily vegetated basins was estimated to increase only slightly or to slightly decrease, depending on estimated changes in temperature and plant water-use efficiency. If total evapotranspiration increases more than these values, total runoff will decrease.

[c]$R_1/R_0$ is the ratio of runoff after a climatic change to the present runoff. Runoff in the Sudd and along the arid stretches of the Nile is conservatively assumed to remain constant, although increases in evapotranspiration in these regions are likely to decrease runoff.

decreased approximately 70% because of a 20% reduction in precipitation. For the Nile Basin as a whole, runoff decreased 33%, making the conservative assumption that there is no significant change in evaporative losses from the region of the Sudd and between Aswan and the Mediterranean. In fact, evaporation in this region is likely to increase, further decreasing overall runoff in the Nile.

The results of this simple model should be considered illustrative rather than predictive, and using data from other GCMs is likely to produce quite different results. Such a simple annual water balance may overestimate the sensitivity of runoff to changes in precipitation in the highlands of Ethiopia, although if precipitation drops, severe decreases in runoff are still likely.

Another factor left out of this type of assessment is the possibility of major shifts in seasonal runoff. In particular, recent hydrologic modelling shows that one of the most likely consequences of greenhouse warming will be a substantial shift in the timing and magnitude of snowmelt runoff in high-altitude river basins with snowfall and snowmelt, such as the Blue Nile. As global temperatures increase, the snowline in such basins will also rise, increasing the winter ratio of rain to snow. This effect, coupled with an earlier and faster snowmelt season, will shorten and intensify seasonal runoff. In the Blue Nile the snowmelt season could be shortened by as much as a month, and peak flows could be substantially higher. Work is underway at the Pacific Institute to try to quantify these changes.

Ultimately, if realistic estimates of changes in regional water availability are to be calculated, a number of advances are needed. In order to be valuable to water resource planners, regional hydrologic assessments must include:

1. A focus on shorter time scales such as months or weeks, rather than annual or even seasonal averages;
2. The ability to incorporate into regional studies the increasingly detailed assessments of changes produced by the large-scale climate models;
3. The ability to produce information on hydrologically important variables, such as changes in runoff and available soil moisture; and
4. The ability to incorporate snowfall and snowmelt, vegetation changes, topography, soil characteristics, natural and artificial storage, and other regional complexities.

Nevertheless, despite the simple nature of the approach used here, this first-order result leads to a strong conclusion: future climatic changes that lead to large changes in water availability must be considered a distinct possibility, and such changes should be incorporated into both planning the physical management of the Nile Basin and into political discussions over equitable sharing of the Nile waters.

## 12.7 Discussion

The challenge of global climatic change makes multinational coordination in the use of shared freshwater resources a necessity. If we wish to avoid future conflicts over limited resources, national self-interest and self-preservation requires that instead of controlling and restricting access to common resources, we must develop mechanisms for the equitable sharing of those resources. While some effort has been made to develop such a management regime in the Nile Basin, the present scheme for managing the Nile is likely to prove inadequate under greenhouse climatic changes.

The fate of the Nile is of immense importance to both Egypt and the Sudan, where it presently sustains a population of 75 million people. Yet the territory of the Nile is shared by nine nations, all of which could influence the flow and quality of the river. Each Nile state, particularly Egypt and the Sudan, assumes that the Nile is the one reliable element in the increasingly complicated challenges that face them. The risks and implications of future climatic changes threaten this assumption.

We cannot yet predict with confidence the nature of future climatic changes in the Nile Basin, but there are indications that such changes will be significant and possibly severe. Indeed, even slight decreases in water availability would place great strains on downstream users. At the same time, the possibility of increases in runoff during the snowmelt season raises the spectre of increased frequency of severe flood events, such as experienced in 1988.

If open conflict is to be avoided over access to the Nile Basin, with or without climatic changes, there need to be modifications in the way the Nile is presently managed. In particular, there is a need to strengthen the Permanent Joint Technical Commission, set up under the 1959 Nile River Agreement between Egypt and the Sudan, and to expand this Commission to include representatives of the other basin states. The Commission should explore the possibility of incorporating more detailed mechanisms for resolving disputes over shortages or severe flood events, both of which are possible outcomes of greenhouse climatic changes. In particular, in times of shortage, some way of changing allocations from a fixed to a proportional flow may be necessary.

The alternative is to do nothing and to assume that the risks of the current management arrangements in the Nile are smaller than the problems that might arise if an attempt to develop a better regime is made. This argument has two serious flaws. First, even in the absence of concern over climatic changes, there are already serious questions about the adequacy of the 1959 Agreement and current management arrangements. This alone suggests the need for a more comprehensive agreement. Second, the magnitude of the likely climatic changes could dwarf the recent natural variations that already have severe societal effects. Unless an effort is made to develop non-confrontational solutions to water scarcity and multinational allocations in the Nile Basin, eventual conflict seems unavoidable.

# References

Budyko, M. I. and Izrael, Y. (1987) *Anthropogenie Izmenenie Klimata (Anthropogenic Climatic Changes)*, Gidrometeoizdat, Leningrad (in Russian).

Butzer, K. W. (1971) *Environment and Archaeology*, 2nd edn, Methuen, Chicago.

Folland, C. K. and Palmer, T. N. (1985) 'Sahel rainfall and worldwide sea temperatures, 1901–1985', *Nature*, 17 April, 602–607.

Gleick, P. H. (1989a) 'Climate change and international politics: problems facing developing countries', *Ambio*, **18**, No. 6. 333–339.

Gleick, P. H. (1989b) 'Climate change, hydrology, and water resources', *Review of Geophysics*, **27**, No. 3.

Goldenman, G. (1989) *International River Agreements in the Context of Climatic Changes*, Pacific Institute for Studies in Development, Environment, and Security, Berkeley, California.

Ross, M. (1987) 'Parched Egypt watches anxiously as waters behind Aswan Dam recede', *Los Angeles Times*, 26 December.

Shahin, M. (1985) *Hydrology of the Nile Basin*, Elsevier Science, Amsterdam.

United Nations (1978) *Registry of International Rivers*, Pergamon Press, Oxford.

Waterbury, J. (1979) *Hydropolitics of the Nile Valley*, Syracuse University Press, New York.

Wickens, G. (1975) 'Changes in the climate and vegetation of the Sudan since 20 000 BP', in *Proceedings 8th Plenary Session AETFAT*, Geneva.

Wigley, T. M. L. and Jones, P. D. (1985) 'Influences of precipitation changes and direct $CO_2$ effects on streamflow', *Nature*, **314** 149–152.

World Resources Institute (1988) *World Resources 1988–89*, Washington, DC.

World Meteorologicial Organization (1987) *Water Resources and Climatic Change: Sensitivity of Water-Resources Systems to Climate Change and Variability*, WCAP-4; WMO/TD-No. 247.

# 13 A STUDY OF THE CLIMATIC VARIABILITY IN THE EASTERN MEDITERRANEAN DURING THE FIRST HALF OF THE TWENTIETH CENTURY

C. C. Repapis and H. T. Mantis
*Academy of Athens, Greece*

D. A. Metaxas and Ph. Kandilis
*University of Athens, Greece*

## 13.1 Introduction

The climate of the earth is continuously being changed by external and internal systems. In recent years a global warming of the air in the lower atmosphere and the oceanic surface layer has been detected. There has been concern that this warming is due partly to the increased clarity of the atmosphere (which recovered from volcanic activity at the beginning of this century) and partly to enhancement of the greenhouse effect (Jones *et al.*, 1986; Budyko, 1988).

Extrapolation by mathematical models of human activities and development indicates a doubling of greenhouse gas concentrations before the middle of the next century, resulting in an increase in the global surface air temperature by approximately 2–3°C. The resulting climate changes will differ among regions, and variations in the forecasts of regional climatic responses are large. It is of great interest therefore to study past regional climatic trends to assess model results for those in the future.

Repapis and Philadras (1988) have reported that warming in the Eastern Mediterranean region during the first half of this century was greater than the hemispheric average, despite the fact that the largest contribution to the latter came from temperature increases at higher latitudes (e.g. Budyko, 1988). In this chapter we report the results of a study of the long-term meteorological records at four stations in the Eastern Mediterranean to determine the homogeneity and temporal structure of climatic change in this region.

## 13.2 Data

Mean monthly temperature and precipitation records at five locations (Figure 13.1) were used in this study: Athens (Greece); Cairo and Alexandria (Egypt); Jerusalem (Israel); Nicosia (Cyprus). The data for Athens were taken from the bulletin of the National Observatory of Athens (1931–1987) and from Arseni-Papadimitriou (1973) while those for Nicosia were provided by the National Meteorological Service of Cyprus. The observation sites at Athens and Nicosia date from the beginning of this century.

The data for Cairo, Alexandria and Jerusalem were taken from the World Weather Records (WWR) and the Monthly Climatic Data for the World (Smithsonian Institution, 1927, 1934, 1947; US Weather Bureau, 1959–1987). The Cairo record after 1904 is continuous at the Helwan site. We have extended the monthly record of Helwan to 1902 using data from Abassia normalized by coefficients obtained from the linear regression of Helwan versus Abassia monthly mean temperatures for the 19-year (1904–1922) interval of overlapping records (the time series of these two stations are shown in the graphs in Figure 13.2). There have been several changes at the Jerusalem observation site which may cause some concern regarding the homogeneity of this record. Homogenized precipitation time series for Jerusalem is published in World Weather Records (US Weather

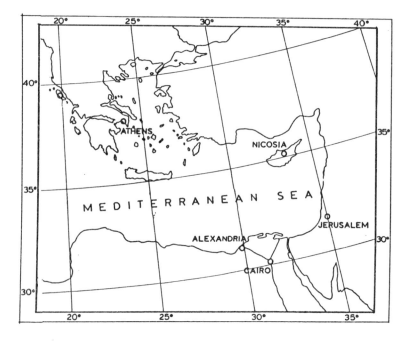

Figure 13.1

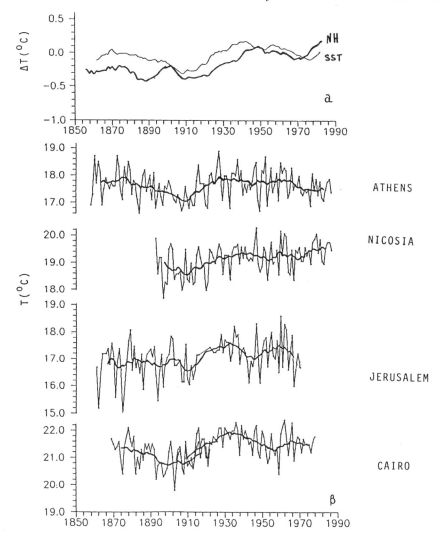

**Figure 13.2**

Bureau, 1966) while the air temperature record for Jerusalem has been estimated by Striem (1974) as homogeneous.

The annual mean air temperature time series for the northern hemisphere were taken from Jones *et al.* (1986) and the sea surface temperature time series from Folland *et al.* (1984). In this chapter the long-term trends are presented and emphasis has been given to a comparison of the characteristics of the climate in the Eastern Mediterranean for a 12-year interval during the colder period at the beginning of this century (1902–1913) to that in the warmer period near the middle (1925–1939). Selection of the boundaries of the warm and cold periods is somewhat arbitrary, but is based in part on the availability of data for these intervals.

The annual mean air temperature values for each station and for each period are shown in Table 13.1.

## 13.3 Discussion and results

### 13.3.1 Air temperature analysis

The air temperature time series for the whole available record in the four stations are shown in Figure 13.2(b). The uniformity of temperature trends in the Eastern Mediterranean is clearly seen. The 10-year running means of the air temperature variations for the Northern Hemisphere (NH) and of the sea surface temperature (SST) are also given for a perspective (Figure 13.2(a).

The annual temperatures for each station and for each period (cold–warm) was obtained by a Fourier analysis of mean monthly temperature using three harmonics. The smoothed curves are illustrated in Figure 13.3 and observed mean monthly temperatures are plotted to show the degree of smoothing by the harmonic filter. The change in mean temperature is almost the same magnitude for all stations in all four seasons. An exception is for February, when it is less; at Athens and Nicosia the observed mean February temperatures in the cold period are slightly higher than in the warm one.

There is always concern that a warming trend seen in the long-term record at an observation station in a city is the result of urbanization. However, urban growth in all four cities has been most rapid in the latter part of this century, during which time there has been a decline in temperature at most stations. Therefore we have made an analysis of the annual course of temperature for Athens for the 15-year period 1968–1982 (termed the control period). The graph of the annual course of temperature in the control period is compared with that for the warm and cold periods in Figure 13.4. While the mean temperature in the warm period is higher than in the control period this variation only occurs in the warmer half of the year. One might speculate that the behaviour of the colder half year could be an urbanization effect (urban heat island).

Table 13.1

| | Mean air temperature (°C) | | | |
|---|---|---|---|---|
| | *1902–1913* | *1914–1924* | *1925–1939* | *1968–1982* |
| Athens | 17.10 | 17.61 | 17.96 | 17.54 |
| Nicosia | 18.63 | 18.97 | 19.29 | |
| Jerusalem | 16.73 | 17.17 | 17.56 | |
| Cairo | 20.89 | 21.52 | 21.86 | |

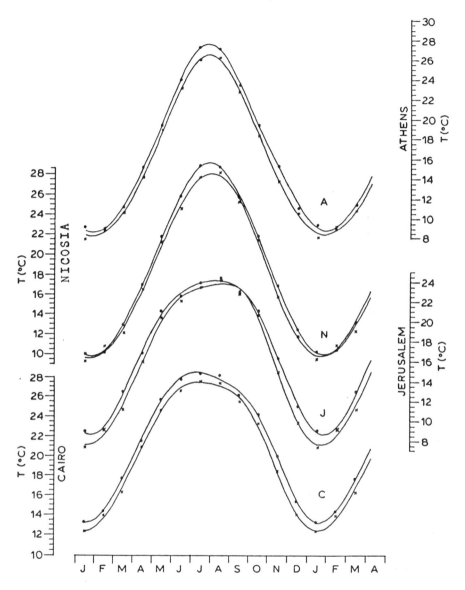

**Figure 13.3**

### 13.3.2 *Precipitation analysis*

In the Mediterranean climate, significant precipitation amounts occur only in the colder eight months of the year and then usually under conditions of strong convection. Precipitation records are therefore unusually erratic, and determination of significant trends in this region is difficult. The time series of the precipitation amounts for the four stations and the 10-year

ATHENS

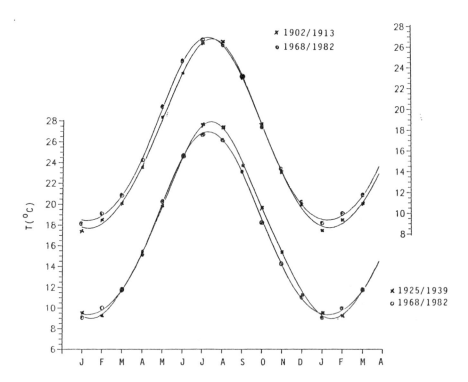

**Figure 13.4**

running means are shown in Figure 13.5. In this figure the Alexandria time series is also shown, as we thought that the very small rainfall amounts in Cairo could obscure any possible trend in that area. The trends are not similar in all the stations. Athens and Nicosia – with long-term means 400 mm and 350 mm, respectively – show a minimum in the rainfall amounts around the turn of this century, while Jerusalem (mean 600 mm) and both Alexandria (mean 190 mm) and Cairo (mean 30 mm) display minima in the 1930s. Mean monthly precipitations for cold, warm periods are compared in Figure 13.6. Only at Jerusalem and Alexandria is there a significant difference in the precipitation regime during these two periods; namely, the cold period has 25–30% higher precipitation than the warm one. In Nicosia the cold period is insignificantly wetter while in Athens the warm period is insignificantly wetter.

The frequency distribution analysis of annual rainfall amounts at Athens reveals that during the warm interval (1925–1939) there is a below-average frequency of low-precipitation years. At Jerusalem, however, during the same interval there is an above-average frequency of low-precipitation years. Striem (1979) has pointed out that cold rainy seasons in Jerusalem are associated with above-normal precipitation amounts.

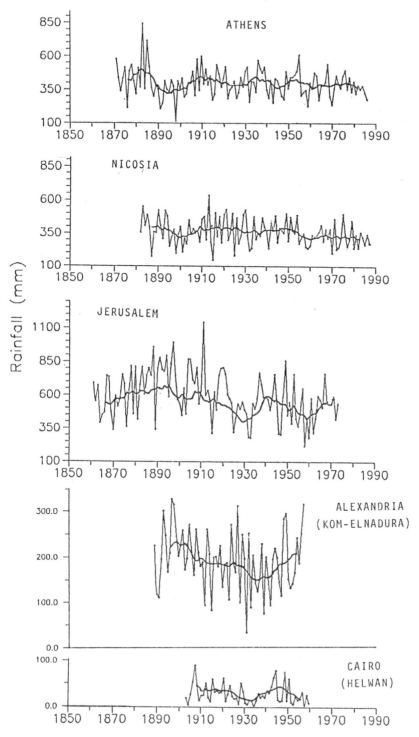

**Figure 13.5**

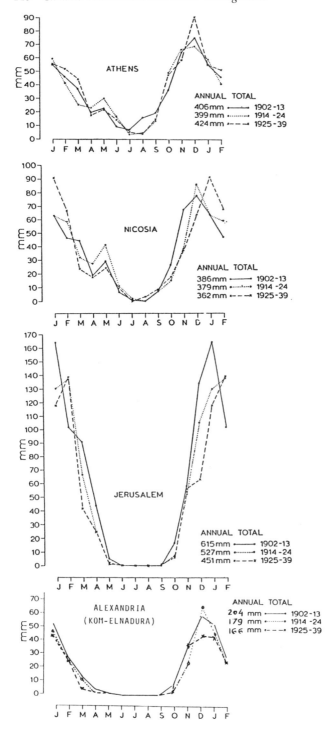

**Figure 13.6**

## 13.4 Conclusions

1. Temperature trends in the first half of this century are similar at all four stations as determined both by annual mean temperatures and by trends in the annual temperature. The representativeness of the trends is supported by the homogeneity of the shorter-term temperature fluctuations. No significant difference in temperature variability was found for the warm and cold periods.
2. The precipitation regime is much more complicated. The rainfall time series for Athens and Nicosia display a minimum at the turn of the century while for Jerusalem and Alexandria the minimum occurs during the 1930s. Although we are aware that the rainfall records are erratic (especially for short periods), we can say that the cold interval of 1902–1913 is significantly wetter than the warm one of 1925–1939 in Jerusalem and Alexandria, while in Athens the warm interval is slightly wetter than the cold one.

## Acknowledgments

This work was sponsored by EEC contracts EV 4C-0027/28 – GR(TT).

## References

Arseni-Papadimitriou, A. (1973) 'On the annual variation of air temperature in Athens', *Sci. Annals Fac. Phys. and Mathem.*, **13**, 325–345.

Budyko, M. I. (1988) 'Climate of the end of the 20th century', *Soviet Meteorology and Hydrology*, **10**, 5–24.

Folland, C. K., Parker, D. E. and Kates, F. E. (1984) 'Worldwide marine temperature fluctuations, 1856–1981', *Nature*, **310**, 670–673.

Jones, P. D., Raper, S. C. B., Bradley, R. S., Diaz, H. F., Kelly, P. M. and Wigley, T. M. L., (1986) 'Northern Hemisphere surface air temperature variations, 1851–1984', *J. Clim. Appl. Meteor.*, **25**, 161–179.

Repapis, C. and Philadras, K. (1988) 'A note on the air temperature trends of the last 100 years as evidenced in the Eastern Mediterranean time series', *Theor. Appl. Clim.*, **39**, 93–97.

Smithsonian Institution (1927, 1935, 1947) *World Weather Records: Smithsonian Inst. Miscellaneous Collections*, Vols 79, 90 and 104, Washington, DC.

Striem, H. L. (1974) 'The mutual independence of climatological seasons, as reflected by temperatures at Jerusalem 1861–1960', *Israel J. Earth Sciences*, **23**, 55–62.

Striem, H. L. (1979) 'Some aspects of the relation between monthly temperatures and rainfall, and its use in evaluating earlier climates in the Middle East', *Climatic Change*, **2**, 69–74.

US Weather Bureau (1959–1982) *World Weather Records*, 1941–1950, 1951–1960 (Vols 1–6), 1961–1970 (Vols 1–6), US Department of Commerce, Washington, DC.

# 14 WEATHER-MODIFICATION IMPACT ON IRRIGATION WATER DEMANDS AND MANAGEMENT IN ARID REGIONS

Walid A. Abderrahman and Asfahanullah Khan
*King Fahd University of Petroleum and Minerals, Saudi Arabia*

## 14.1 Introduction

It is well established that an increase in the levels of carbon dioxide ($CO_2$) and other gases such as methane and ozone in the troposphere are responsible for warming the earth through what is known as the greenhouse effect. Some meteorologists suggest that there will be a global increase in air temperature by 1.5–4.5°C during the next 50 years ·if the present trend continues. The increase in air temperature is expected to be accompanied by global changes in other meteorological parameters such as wind speed, sunshine hours, humidity, precipitation and radiation. Complex simulation models are used to forecast these global changes.

Management of water resources depends mainly on water demand and the availability of water. In arid regions such as the Arabian Peninsula, irrigation demands comprise more than 75% of total water demand. These demands are satisfied mostly from groundwater resources.

Any increase in air temperature or change in other meteorological parameters in arid regions will affect the irrigation water demands, growth and yield of the cultivated crops. Assessing the increase in irrigation water demands resulting from a modification of the weather is important for developing future water supply alternatives and water management schemes.

This chapter describes the impact an increase in air temperature will have on irrigation demands, crop yield and water management in three main agricultural areas in the Arabian Peninsula: coastal areas represented by Qatif, oasis areas by Hofuf and central areas by Kharj (Figure 14.1).

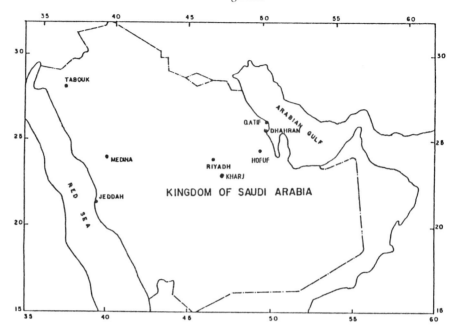

**Figure 14.1**   *Map showing the locations of the study areas in the Arabian Peninsula*

## 14.2 Climate

The Arabian Peninsula is located in the arid region of the world. The average annual precipitation is less than 100 mm in coastal and oasis areas, and ranges from about 100 to 200 mm in the central highlands. In the south-western mountains the average annual rainfall is from about 300 mm to 500 mm. There is a great variation in air temperature between summer and winter and day and night in the three areas. In the summer season the average air temperature ranges between 27° and 37°C in central areas, 32° and 38°C in oasis areas and 31° and 36°C in coastal areas. In the winter season (December to February) the average air temperature ranges between 4° and 20°C in central areas, 8° and 17°C in oasis areas and 11° and 22°C in coastal areas. The average values of relative humidity in summer and winter are about 75% and 65%, respectively, in coastal areas, about 35% and 60%, respectively, in oasis areas and about 15% and 55%, respectively, in central areas. The average annual evaporation values are high in the three areas. They are about 3000 mm, 3500 mm and 3500 mm in coastal, oasis and central areas, respectively.

## 14.3 Methodology

In this chapter the effect of air temperature increase on irrigation demands was assessed assuming no change in other climatic parameters. The effect of a temperature increase from 1°C to 5°C on irrigation requirements was evaluated in three areas in the Arabian Peninsula. The crop irrigation requirement (*IR*) was calculated from the recommended method of Doorenbos and Pruitt (1984) using the following relationship:

$$IR = (ET_{crop} - (P_e + G_e + W_b))/(1 - LR)E_p \tag{14.1}$$

where

$IR$ = crop irrigation requirements from the source of water (mm/day),
$ET_{crop}$ = crop evapotranspiration (mm/day),
$P_e$ = contribution from rainfall (mm/day),
$G_e$ = contribution from shallow groundwater (mm/day),
$W_b$ = contribution from carryover of soil water (mm/day),
$LR$ = leaching requirements, and
$E_p$ = irrigation efficiency.

In equation (14.1) the major impact of temperature increase on the irrigation requirements (*IR*) is the effect on crop evapotranspiration ($ET_{crop}$). Crop evapotranspiration ($ET_{crop}$) was calculated from

$$ET_{crop} = ET_0 \cdot K_c \tag{14.2}$$

where

$ET_0$ = reference crop evapotranspiration (mm/day),

$K_c$ = crop coefficient which is mainly affected by crop characteristics.

Temperature change on $ET_{crop}$ value affects the value of reference crop evapotranspiration ($ET_0$).

Increases in daily values of reference crop evapotranspiration ($ET_0$) were calculated for each month of the year using the modified Penman method (Doorenbos and Pruitt, 1984) from the following relationship:

$$ET_0 = c(W \cdot R_n + (1 - W) \cdot f(u) \cdot (e_a - e_d)) \tag{14.3}$$

| Radiation | Aerodynamic |
| term | term |

where

$W$ = temperature-related weighting factor,
$R_n$ = net radiation in equivalent evaporation (mm/day),
$f(u)$ = wind-related function,
$(e_a - e_d)$ = difference between the saturation vapour pressure at mean air temperature and the mean actual vapour pressure of the air (both in mbar),
$c$ = adjustment factor to compensate for the effect of day and night weather conditions.

Several methods were examined to calculate the changes in $ET_0$ values (e.g. Blaney Criddle, radiation, Penman, modified Penman, pan evaporation, and Jensen–Haise). The modified Penman method was found the best for predicting the $ET_0$ values when compared with measured $ET_0$ values (KFUPM/RI, 1986).

A special computer program was used to calculate the change in values of reference crop evapotranspiration ($ET_0$) and irrigation requirement ($IR$) with increases in temperature by 1°, 2°, 3°, 4° and 5°C in the three geographical areas. Meteorological data of the Ministry of Agriculture and Water (1987) for Qatif (1976–1984), Hofuf (1969–1987) and Kharj (1976–1984) were used in the calculation of $ET_0$ values. These included temperature, humidity, solar radiation, sunshine hours and wind speed. The increase in $IR$ value is equal to 1.54 times the increase in $ET_0$ value, as the average irrigation efficiency value $E_p$ is considered to be 65% in the peninsula.

If excess water demands are not satisfied, crop yield will be reduced in relation to its yields response factor ($K_y$). The effect of water shortages on the yield of three main crops – alfalfa, tomatoes and date palms – were calculated. The method recommended by Doorenbos and Kassam (1986) was used from the following empirical relationship:

$$1 - Y_a/Y_m = K_y(1 - ET_a/ET_m) \tag{14.4}$$

where

$Y_a$ = actual harvested yield (ton/ha),
$Y_m$ = maximum harvested yield (ton/ha),
$K_y$ = yield response factor for a crop type grown in certain location and soil type,
$ET_a$ = actual evapotranspiration (mm/day),
$ET_m$ = maximum evapotranspiration (mm/day).

The $K_y$ values used for alfalfa, tomatoes and date palms are shown in Table 14.1.

**Table 14.1** Yield response factor of alfalfa, tomatoes and date palms (from Doorenbos and Kassam, 1986; HARC, 1976)

| Crop | $K_y$ |
| --- | --- |
| Alfalfa | 0.3–0.42[a] |
| Tomatoes | 1.05 |
| Date palms | 0.65–1[b] |

[a]This is an empirical value taken from HARC (1976); $K_y$ = 0.3 at water shortages up to 30%; and 0.42 at 40% water shortage.
[b]0.65 during the months February to May; and 1 during the remainder of year.

## 14.4 Results and discussion

*14.4.1 Irrigation demands*

Calculated increases in the values of $ET_0$ and $IR$ in the three geographical areas are plotted in Figures 14.2–14.7. In general, the rise in temperature resulted in increasing the $ET_0$ and $IR$ values in the three areas by different ratios. The increase in $ET_0$ and $IR$ values during the winter season (November–March) were greater than in the summer and autumn seasons (April–October).

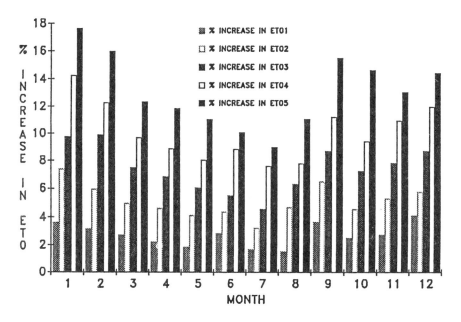

**Figure 14.2** *Percentage increase in* $ET_0$ *values with the expected rise in temperature in a coastal area (Qatif)*

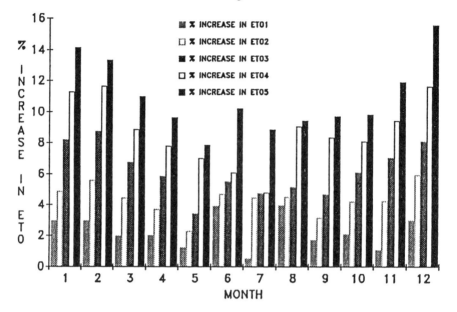

**Figure 14.3**  *Percentage increase in* $ET_0$ *values with the expected rise in temperature in an oasis area (Hofuf)*

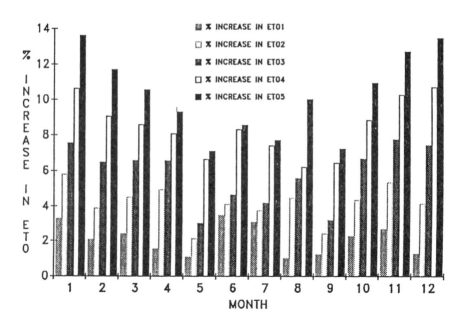

**Figure 14.4**  *Percentage increase in* $ET_0$ *values with the expected rise in temperature in a central area (Kharj)*

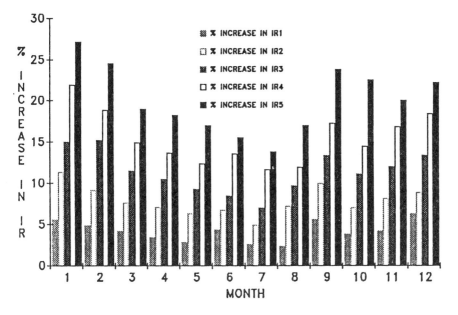

**Figure 14.5**   *Percentage increase in* IR *values with the expected rise in temperature in a coastal area (Qatif)*

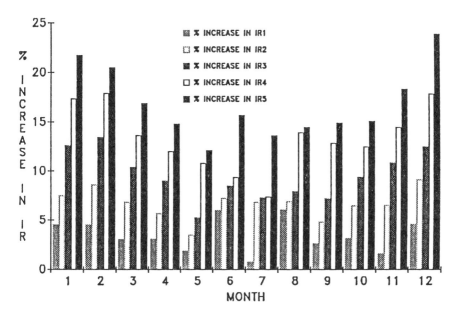

**Figure 14.6**   *Percentage increase in* IR *values with the expected rise in temperature in an oasis area (Hofuf)*

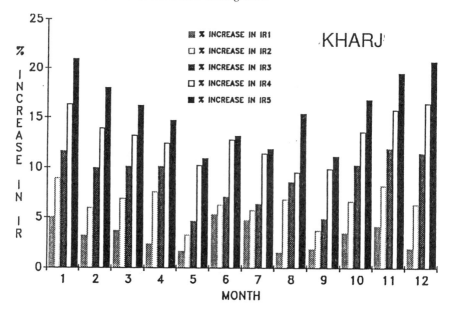

**Figure 14.7**   *Percentage increase in* IR *values with the expected rise in tempera-ture in a central area (Kharj)*

In the coastal area (Qatif), the percentage rise in $ET_0$ with a 1°C increase in temperature ranged from 1.5% in August to 4.2% in December. With a 5°C rise, $ET_0$ increased from 9% in July to 17.7% in January (Figure 14.2). In the oasis area (Hofuf), a 1°C rise in temperature increased $ET_0$ from 0.5% in July to 3% in December. A 5°C rise resulted in an $ET_0$ increase ranging from 7.9% in May to 15.5% in December (Figure 14.3). In the central area (Kharj), a 1°C rise in temperature increased $ET_0$ by values ranging from 1% in August to 3.3% in January (Figure 14.4).

The maximum increases in *IR* through a 1°C increase in temperature were 1.1%, 0.4%, and 0.9% in coastal, oasis and central areas, respectively (Figures 14.5–14.7). The maximum increases in *IR* by a 5°C rise in temperature were 27.2%, 23.9% and 21.0% in coastal, oasis and central areas, respectively. Generally, the rise in $ET_0$ and *IR* values in coastal areas > oasis areas > central areas at any increase in temperature.

### 14.4.2 *Yield reduction*

Possible yield reductions of alfalfa, tomatoes and date palms were calcu-lated in the three geographical areas assuming that the increases in water requirements were not satisfied (Table 14.2). The calculated yield reduc-tions for the three crops were in coastal area > oasis area > central area. The yield losses in the coastal area with a 1°C increase in temperature were 0.9% in alfalfa, 3.2% in tomatoes and 2.0–3.0% in date palms. Yield losses

**Table 14.2** The calculated yield reduction in alfalfa, tomatoes and date palms at different levels of water shortages

| Water shortage (%) | Alfalfa | Tomatoes | Date palms |
|---|---|---|---|
| 1.0 | 0.30 | 1.0 | $0.6^a$–$1.0^b$ |
| 1.5 | 0.45 | 1.5 | 0.9–1.5 |
| 2.0 | 0.60 | 2.1 | 1.3–2.0 |
| 3.0 | 0.90 | 3.1 | 1.9–3.0 |
| 4.0 | 1.20 | 4.2 | 2.6–4.0 |
| 5.0 | 1.50 | 5.2 | 3.3–5.0 |
| 6.0 | 1.80 | 6.3 | 3.9–6.0 |
| 7.0 | 2.10 | 7.3 | 4.5–7.0 |
| 8.0 | 2.40 | 8.4 | 5.2–8.0 |
| 9.0 | 2.70 | 9.4 | 5.8–9.0 |
| 10.0 | 3.00 | 10.5 | 6.0–10.0 |
| 15.0 | 4.50 | 15.7 | 9.8–15.0 |
| 17.0 | 5.10 | 17.8 | 11.1–17.0 |
| 20.0 | 6.00 | 21.0 | 13.0–20.0 |
| 25.0 | 7.50 | 26.2 | 16.2–25.0 |
| 30.0 | 11.00 | 31.5 | 19.0–30.0 |
| 35.0 | 14.70 | 36.6 | 22.7–35.0 |
| 40.0 | 16.80 | 42.0 | 26.0–40.0 |

[a]Yield reduction if the shortage occurs between February and May.
[b]Yield reduction if the shortage occurs between June and January.

due to a 5°C increase in temperature were 3.9% in alfalfa, 13.7% in tomatoes and 8.5–13.1% in date palms. Yield losses at any level of water shortage were tomatoes > date palms > alfalfa.

There is an indirect effect caused by the water shortages on the salinity of the root zone. With increasing crop water requirements, the plants consume water from the leaching fraction. This will increase soil salinity, causing more yield losses.

### 14.4.3 Water management problems

Under field conditions in the Arabian Peninsula the groundwater is used mostly to cope with water shortages. If shortages during the winter season are overcome, this affects the water level and well productivity in the summer season. Consequently, water shortages and yield losses are expected to be greater during that season.

A long-term increase in groundwater withdrawal will significantly lower water levels in the aquifer systems. This will lead to deepening the irrigation wells and increasing the costs of pumping. The quality of water may deteriorate with drops in water levels. To compensate for a possible increase in irrigation demands, improvements in irrigation operation and efficiency are essential. These results indicate clearly that it is very important to eliminate the causes of the greenhouse effect by maintaining the air temperature at its natural level. This will avoid an increase in irrigation demands and preserve the water resources.

## 14.5 Conclusions

The possible increase in the earth's temperature by the greenhouse effect by 1–5°C will have a considerable influence on irrigation water demands in arid regions such as the Arabian Peninsula. A rise in temperature from 1° to 5°C could increase irrigation demands by 0.9% to 27% in three agricultural areas within the Peninsula. Compensating for the increase in water demand by withdrawing more water from the aquifer systems will result in lowering the water levels, increasing the cost of pumping and further depleting water resources. If the additional water demands are not satisfied, the crops will suffer from yield reductions according to the yield response factor of each crop. Effective measures should be taken to eliminate the causes of increasing the greenhouse effect to maintain the global temperature at its natural levels and to protect water resources.

## Acknowledgements

The authors are grateful to the Research Institute of King Fahd University of Petroleum and Minerals for providing support to carry out this study.

## References

Doorenbos, J. and Kassam, A. H. (1986) 'Yield response to water', Irrigation and Drainage Paper No. 33, FAO of the United Nations, Rome.
Doorenbos, J. and Pruitt, W. O. (1984) 'Crop water requirements', Irrigation and Drainage Paper No. 24, 2nd edn, FAO of the United Nations, Rome.
Hofuf Agricultural Research Center (1976) *Irrigation Handbook*, Report on the work of Leichtweiss Institute Research Team, Technical University Braunschweig, Publication No. 17, Hofuf, Saudi Arabia.

King Fahd University of Petroleum and Minerals/The Research Institute (KFUPM/RI) (1986) *Modeling and Programming of Al-Hassa Irrigation System*, Unpublished report, Dhahran, Saudi Arabia.

Ministry of Agriculture and Water (1987) *Meteorological Data*, Unpublished data, Hydrology Section, Ministry of Agriculture and Water, Riyadh, Saudi Arabia.

# 15 IMPLICATIONS OF SEA-LEVEL RISE ON THE DEVELOPMENT OF LOW-LYING COASTAL AREAS: THE NEED FOR A QUANTITATIVE APPROACH

C. H. Hulsbergen
*Delft Hydraulics, The Netherlands*

## 15.1 Introduction

Global warming due to the greenhouse effect will probably cause the mean surface level of the oceans to rise between 0.5 m and 2 m by the year 2100. Such an increase will have a considerable impact on society, the consequences of which are about to explored in a number of studies.

A rise in sea level particularly affects low-lying coastal areas, which, in general, exhibit sensitive (thus vulnerable) dynamic equilibrium conditions in terms of coastline morphology, river-mouth behaviour, salinity patterns, drainage, and an entire range of marine, fluvial and subearial biotic factors. The first results of a global inventory concerning the vulnerability of coastal areas to such a rise strongly point towards the developing countries as major problem areas for both physical and socio-economic reasons.

Amid all issues on climate change, the expected sea-level rise is a most pressing reason to limit the further pollution of the atmosphere with greenhouse gases as much as possible. However, even if greenhouse gas emission did cease immediately, thermal inertia effects in the oceans and the land ice would still cause an appreciable rise in sea level due to the global warming of the altered atmosphere.

The accelerated threat of an increase in sea level appears at a time when many other (man-made) problems are already being experienced in the coastal areas of the world, i.e.:

1. Exponential population growth, resulting in competing, conflicting and destructive exploitation of natural resources;
2. Increasing man-made subsidence of coastal areas due to extraction of water, oil and gas;
3. Rapid (man-made) coastline erosion due to a disturbed sediment balance, caused by upstream water management (reservoirs), beach and coral mining, and deteriorating coastal structures.

If considered from a short-term point of view, an accelerated rise in sea level is just one additional factor with which engineers have to deal, in that a further rise 'merely' constitutes a weakly time-dependent hydraulic boundary condition for the design of well-known types of coastal engineering structures. However, when regarded in a wider perspective, the expected rise, through its tenfold acceleration as compared to recent historic (quasi-constant) behaviour, will have the destructive effect of a worldwide tidal surge caused by an earthquake. The elevation of the mean sea level has always, on a human and engineering development scale, served as the ultimate, reliable reference to locate and to design settlements, structures, drainage works and all other infrastructure in the world's coastal areas. Through the greenhouse effect it is now envisaged that the mean sea level itself will rise by one metre during the period of one human generation, thereby not only encroaching upon the world's densest human development areas (i.e. coastal zones) but also jeopardizing the settling areas for many billions of people in the twenty-first century.

How can we deal with the expected rise in sea level? What values are at stake in terms of personal safety, socio-economic damage, natural habitats and human and natural resources in general? What kind of human response can be conceived, from simple abandonment of coastal areas to sophisticated engineering defence measures? How much time is needed and available to define and implement feasible response policies? What criteria may be used in decisions to abandon or protect areas threatened by a sea-level rise? What might be the costs, and who should pay?

These questions have been posed, and preliminary approaches formulated to some of them. It has appeared very difficult to produce straightforward and realistic answers, due to a shortage of basic field data and the involvement of overwhelmingly complicated societal, physical and biotic relationships and processes.

The aim of this chapter is to give a brief survey of the impact and response approaches which have been initiated so far, itemized as follows:

- Sea-level rise projections
- Impact of such a rise on society
- Response policies
- Conclusions

## 15.2 Sea-level rise projections

Despite much research, there does not exist a complete quantitative insight into the main factors which together result in the height of the global sea level. The continuously varying (wave-smoothed) elevation of the sea surface is just one of a very large number of physical processes which form a complicated system of meteorological, hydrological, biological and geophysical functional relationships. Excellent overviews regarding recent study results of past sea-level changes have been carried out by Street-Perrott *et al.* (1983), Van de Plassche (1986), Tooley and Shennan (1987) and Devoy (1987). The interesting aspect of these historic analyses of changing sea level is the possibility of combining the main dynamic factors in mathematical models. These might, after adequate verification and calibration, serve as tools to predict a future sea-level rise under various scenarios and include anthropogenic factors such as climate changes caused by inadvertent atmosphere manipulation (greenhouse effect).

During a recent SCOPE Workshop on Sea Level Rise, Raper *et al.* (1988) examined predictions of a greenhouse-based rise in sea level, especially the assumptions which had been used to arrive at these predictions. Raper *et al.*'s paper consists of four parts. In the first part they review existing analyses of worldwide tide gauge data over the last century, and estimate, after careful consideration of possible error sources, that the sea level has shown an absolute rise of 12 cm ± 12 cm over the last 100 years, implying that there is a very small probability that the global sea level has not changed over this period. Local land level (= gauge level) changes should be added to arrive at trends in local relative sea-level rise. Second, Raper *et al.* evaluate global temperature change studies pertaining to the last 100 years, and conclude that the average global temperature has increased by 0.5°C ± 0.1°C. They state that the 0.5°C warming is consistent with estimates of warming due to increases in greenhouse gases. In the third part of their paper the authors give a quantitative consideration of how a global warming of 0.5°C could change the sea level through thermal expansion of the oceans and by changes in the net mass balance of small glaciers and large ice sheets. Thermal expansion could account for approximately 15–50% of the estimated 10–15 cm rise in sea level, and the contribution of small glaciers (excluding those of Greenland) could be 15–70% of the total rise in sea level. Much less certain are the estimates of Greenland and Antarctica mass balance contributions. In fact, error bars in their present estimates are so large that a negative total figure for sea-level rise cannot entirely be excluded. In summing up the first three parts Raper *et al.* conclude that, despite the great uncertainties, there appears to be reasonable grounds for supporting the postulated positive trend in eustatic sea level, based on temperature-related causes – particularly thermal expansion and small glacial melting.

More importantly, this lends credence to the supposition that greenhouse warming may cause the rate of sea-level rise to increase in the future, which is the subject of the fourth part of their paper. In a discussion of available sea-level rise computations and estimates. Raper *et al.* ascribe their relatively large discrepancies ('best estimates' range from 5 cm to 10 cm per decade, while sea-level rise estimates for the year 2100 are between approximately 0.5 m and 3.5 m) to the imponderable mix of assumptions and methods. They state that there are three major uncertainties in estimating future greenhouse warming: (1) rate of emissions and consequent atmospheric concentrations of greenhouse gases; (2) the sensitivity of climate (mean global temperature) to such increased concentrations; and (3) the damping effect due to the thermal inertia of the oceans. In their own projections for the year 2030, Raper *et al.* include a 'best-estimate' equivalent $CO_2$ concentration of 538 ppmv, reflecting the likely effect of the recent Montreal Protocol to reduce CFC production. Further, they used estimates of the equilibrium change in global mean air temperature for a doubling of $CO_2$ ($\Delta T_{2x}$) as determined from climate models, with a best estimate of 3.0°C, in accordance with recent assessments. Finally, for the oceanic diffusivity coefficient $k$ they adopt a best estimate of 1.0 $cm^2s^{-1}$, which determines the rate of ocean warming.

With this set of assumptions Raper *et al.* made estimates of the 1985–2030 equilibrium and transient 'warming commitments'. Their best guess is that, by the year 2030, we will have committed ourselves to an eventual (equilibrium) global warming of 1.5–3.6°C higher than today. The corresponding transient warming, calculated from 1765, is 1.0–2.2°C warmer than today. This represents a rate of warming some four to ten times faster than experienced over the last 100 years. Starting from this global warming scenario, Raper *et al.* arrive at a 'best-guess' sea-level rise in 1985–2030 of 12–18 cm, of which thermal expansion contributes 6–10 cm, Alpine glaciers 5–8 cm, Greenland 2–3 cm, and Antarctica −1 cm to −3 cm. Further extension of their results to the year 2100 is in progress.

By the end of 1988 WMO and UNEP initiated a worldwide effort to collect and evaluate present knowledge in the field of climate change and its potential impacts on mankind, the so-called Intergovernmental Panel on Climate Change (IPCC). This will result, through activities in three Working Groups and Sub-Groups which organize a number of workshops and seminars, in a state-of-the-art final report which is due by mid-1990, including the latest views on expected sea-level rise.

In conclusion, the rate of absolute sea-level rise will undoubtedly increase due to the greenhouse effect, though this is difficult to predict quantitatively with much exactitude, for well-known reasons. Though estimates can be found in the literature ranging between 0.5 m and 3.5 m for the rise by the year 2100, recent analysis indicates that a range of between 0.5 m and 2.0 m is to be adopted as a more realistic expectation. Local relative sea-level rise may differ considerably from the above-mentioned rate, due to local and regional land-level rise or subsidence, which is not directly related to the greenhouse effect.

# 15.3 Impact of such a rise on society

Every conceivable impact which an accelerated rise in the sea level may have on coastal areas and on human society is known already and has occurred in many locations:

- Flooding of low lands and wetlands
- Devastation of settlements, houses, crops, infrastructure
- Erosion of sandy beaches and dunes
- Breaching of dikes
- Inundation of polders (i.e. reclaimed land)
- Loss of life
- Salination of fresh water reservoirs, lower river branches and aquifers
- Abandonment of inhabited islands
- Disruption of communications
- Blocking of natural drainage, thus causing upstream flooding by rivers
- Shoaling of navigation channels
- Damaging of coastal structures such as harbours, outfalls, inlets, off-shore structures

The 'only' difference – compared to today's situation – which will be seen with an increasing rate of sea-level rise is that all these impacts will occur:

- More frequently
- Further inland
- Over larger coastal areas
- In more countries
- To communities which until now have not had to cope with these problems

This causes an additional burden in terms of increased local, regional, national and global stress on the natural and human resources in the coastal zone, precisely in those areas which are already (or will be in the near future, due to increasing population) overstressed. The overall average effect on society will be comparable to an increased tax level without having access to the revenues, and, moreover, having fewer resources from which to pay the tax. The average global standard of living will fall as sea level increases.

The question of the general, average impact of sea-level rise is not, however, as interesting as addressing the specific vulnerability of particular areas or nations. Without explaining the concept of vulnerability in detail, it will be clear that there are large differences in vulnerability among the world's coastal nations. In this respect, two main aspects may be discerned, a 'damage' component ('the impact') and a 'response' component. Both make up the effective vulnerability of a nation *vis-à-vis* an accelerated sea-level rise. For example, the Netherlands and Egypt have roughly the same vulnerability in the purely physical 'impact' sense. However, the Netherlands, with its higher income and an elaborate coastal defence culture, is less vulnerable than Egypt, since coping with the same problems is a relatively major and unknown undertaking for the latter.

This concept of vulnerability has been further elaborated in a recent global inventory study for UNEP, addressing various criteria for assessing vulnerability (Delft Hydraulics, 1989). This identifies the following among the group of most vulnerable countries: Bangladesh, Egypt, Indonesia, Maldives, Mozambique, Pakistan, Senegal, Surinam, Thailand and the Gambia. However, the study was seriously hampered in its quantitative vulnerability assessment by the lack of basic data pertaining to the coastal areas in the more vulnerable and less-developed countries. Site-specific research and data collection is deemed a first requirement before a reasonable quantitative estimate of the various vulnerability elements can be made.

Such research has begun for the Netherlands (Delft Hydraulics, 1988) and for the Maldives (Hulsbergen and Schröder, 1989) while in other countries such as Bangladesh and Egypt a start will soon be made with detailed investigations to quantitatively assess the impact of sea-level rise on their societies.

## 15.4 Response policies

Response options fall broadly in the range of abandon, retreat, adapt, defend, reclaim, and attack. In this context, 'attack' means, for example, closure of estuaries, thereby shifting the defence line towards a more seaward position. Engineering implications for response options have been described for the United States (National Research Council, 1987).

How can we select one or more of these options, and decide whether the large investments which will be involved are justified? If no decisions are taken because of the complexity of the problems, what values are threatened in the long run? These questions have been addressed by Baarse and Rijsberman (1987) during a Workshop on the Impact of Sea Level Rise on Society (ISOS), and some of their views with respect to the 'ISOS-model' for policy analysis are given below.

As a sea-level rise may change the spatial dimensions and physical appearance of man's living environment, it affects all interests associated with the functions and values of this environment (e.g. housing, food production, industrial production, infrastructural and other man-made facilities, ecological performance, etc.). The minimization of harmful effects may bring about a variety of trade-offs, which will usually involve investment patterns, on the one hand, and all kinds of socio-economic and environmental effects, on the other (loss of facilities and capital goods, food production capacity, safety, disruption of natural systems, etc.). One of the implications is that the problem and its effects need to be studied by a multidisciplinary team, including, among others, physical scientists, economists, ecologists, civil engineers and sociologists.

In general, a large variety of possible measures/strategies can be conceived. In terms of coastal defence options, many possibilities emerge when taking into account the specifications of, for example, type of structure, location, materials used, dimensions, phasing in time, etc. The same is true when considering options related to a change in land use, or remedial measures to reduce some of the adverse effects with respect to, for example, water management, salinization, port structures and operations. Clearly, numerous combinations of such measures may be possible. The complexity of the analysis is further emphasized by the dynamics of the processes involved. This applies to the response of natural systems and processes (e.g. the ecological system or morphological processes) as well as to the response of people in general (households) and specific interest groups such as owners of threatened land and facilities (farmers, industries), people who are to be resettled, nature conservancy groups, etc.

The decision-making structures will typically be complex. The problem of sea-level rise will affect all governmental levels and departments, and is likely to activate many interest groups. It also involves international aspects (mostly from the viewpoint of developing common approaches and sharing information). An important complication emerges from the fact that decision makers in general tend to have a relatively short-term view, which is clearly inadequate for the kind of problems and solutions to be considered.

Not only is the problem of sea-level rise surrounded by uncertainty due to, *inter alia*, lack of knowledge of many of the complex processes related to the causes and effects of such a rise, there is also uncertainty about future developments. The nature of the problem requires the introduction of a relatively long time-horizon. This causes large uncertainties about future developments, not only with respect to the phenomenon of sea-level rise itself but also regarding other natural conditions, demographic and socio-economic developments (activity levels, income situation), land use, etc.

The problem of sea-level rise requires a well-structured approach that fits the above characteristics and needs. It is felt that the concept of policy analysis will meet these requirements.

## 15.5 Policy analysis

A policy analysis basically aims at providing an overview of the relevant effects of alternative strategies in support of decision making. The key elements are:

- Relevant effects
- Alternative strategies
- Decision making

The analysis should facilitate the decision-making process, i.e. the information produced should be useful to decision makers. At least three implications follow from this. First, the information should be easily *understood* and interpreted. Second, it should be complete, or at least should take into account all important aspects that are *relevant* for the decision-making process. Third, the results of the analysis should be *acceptable* to the decision makers.

To meet the first requirement, the information collected, computed and presented should be expressed in logical and meaningful units, with well-structured tables or graphs, provided with adequate explanation. It is important that only a limited amount of information is provided at any one time.

To satisfy the second requirement, no information that may be relevant in the decision-making process should be omitted, even if the assessment of certain effects is only qualitative. As uncertainty usually plays a major role, the risks of being wrong should be taken into account explicitly in presenting the effects of alternative options. In addition, the information presented should consider the most relevant trade-offs between, for example, costs, environmental effects, safety, housing conditions, food supply and income situation. The acceptability of the results can be greatly enhanced if one is very explicit about the way the impacts were assessed, i.e. clear insight should be provided into the system assumptions and scenarios used, data sources and data-processing procedures, mathematical models and computational techniques, calibration and verification.

A policy analysis aims at providing information to support decision making. Such information pertains to the effects or impacts of alternative strategies or policies. Evaluation and decision making are based on a comparison of such effects and, for ease and clarity, the effects of alternative strategies should be expressed quantitatively as far as possible.

A comparison of actual numbers provides the decision maker with a clear insight into the relative effects on important impact categories of alternative strategies. It allows him to consider different combinations of measures and strategies to suit his set of value judgements. It also allows very explicit trade-offs to be made against different types of impacts.

However, the quantification of effects may have severe limitations. In certain situations these limitations may be such that an actual quantification would be no longer useful in the evaluation process. This may be the case if the quantitative assessment is uncertain, unreliable or even unrealistic. Another potential problem is that the desire to quantify effects may limit the scope of the analysis beyond reality (for example, effects that cannot be quantified are simply omitted).

The conclusion is that a quantification of impacts is inevitable to support the decision-making process in complex problem fields. Yet one should not so much follow a simple quantitative approach but rather one that aims at providing a complete and unbiased overview of all possible consequences of alternative strategies. A sound policy analysis should therefore take into account the following rules:

1. Boundary conditions, starting points and systems assumptions should be stated explicitly.
2. A complete overview of relevant effects, whether quantified or not, should be provided.
3. An adequate insight into the consequences of major uncertainties should be provided by applying sensitivity analysis and/or using scenarios.

A computational framework will be very suitable for generating the quantitative information required about the problem of sea-level rise, provided that the above rules are taken into account. The main functions of such a computational framework, referred to as the 'ISOS model', are:

1. To integrate or contrast the results of detailed studies on the specific effects of a sea-level rise in order to detect inconsistencies and/or the need for additional information;
2. To investigate the effects of specific individual measures taken to prevent a sea-level rise, reduce damage resulting from a rise, or alleviate its negative effects in order to 'design' these measures in such a way as to generate maximum beneficial effect at minimum cost;
3. To analyse the effects of uncertainty in the various components of the analysis, of the causes and effects of the sea-level rise itself, and the effects of measures, mainly through analysis of a set of scenarios;
4. To evaluate a set of alternative strategies to deal with a sea-level rise for a given study area and time horizon to generate information that can be used by decision makers to select the most appropriate strategy for implementation.

In view of the nature of the problem and related phenomena, and the objectives listed above, the ISOS model should meet a number of different requirements, the most important of which are outlined below:

1. The model should provide an overview of all impacts of sea-level rise (economic, public health, environmental/ecological, social, and administrative or institutional) in all relevant zones (seas and oceans, estuaries/wetlands, shoreline, lower rivers, and land) over the adopted time horizon. In other words, it should have a multidisciplinary set-up, and should take into account spatial and temporal effects.
2. The model will serve as a tool for improving communication between disciplines. This means that differences in the nature of the impact assessment for the various sectors (for example, a quantitative versus a more qualitative approach) and differences in accuracy will have to be dealt with appropriately.
3. The model should facilitate interaction between decision makers. To do this, it should be used interactively in a conversational mode and be easy to use by non-experts and the results should be presented in comprehensive overviews (e.g. in summary tables and graphs).

4. The model should be able to produce results which reflect the costs and effects of possible measures or strategies and to take into account relevant scenarios (e.g. those related to the extent of sea-level rise, economic developments, etc.).
5. The model should be able to deal with trade-offs in time, i.e. investments made previously to anticipate future problems, given time delays.

Obviously, it will not be possible to quantify all impacts such as the effects on public health or on the environment. On the other hand, the rationale of a policy analysis is that it is helpful to decision makers in that it provides as much quantified and properly organized information as is reasonably possible. This significantly affects the design of the ISOS model. It implies that the model will be based on a mix of quantitative and qualitative information, and that its final results or outputs will have to be in the form of tables or graphs that summarize a set of effects rather a single number or ranking.

It is clear that the relationships between the processes, phenomena and effects involved are, in most instances, of a quite complex nature. If a more or less detailed description of these relationships (if available) can be included in the model while, at the same time, meeting the above requirements, the model would probably grow beyond reasonable proportions and become unmanageable. Therefore the ISOS model should be considered as a 'simplified overview' in which each equation or parameter reflects the result of a more sophisticated model, a more detailed analysis, expert knowledge or good judgement. As such, the 'ideal' ISOS model would contain all the available knowledge about all relevant aspects in a highly condensed and simplified form. It could then be used and understood by the people actively involved in (the preparation of) decision making.

The structure and use of a model that will meet the requirements outlined above have been elaborated by Baarse and Rijsberman (1987). A model of this type has proved to be a fruitful approach in formulating and comparing coastal defence options for the protection of the Netherlands (Delft Hydraulics, 1988).

## 15.6 Conclusions

1. Some impacts of sea-level rise are very well known, as witnessed by disasters on shorelines in perpetual conflict with waves, floods and salt water intrusion.
2. An accelerated rise of the level of the seas and oceans will increase the frequency and intensity of these impacts.
3. National resources will increasingly be affected, in some cases jeopardizing nations which are particularly vulnerable.

4. Absolute global sea level will rise at an accelerating rate due to the greenhouse effect.
5. By the year 2100 the sea level is expected to be 0.5–2.0 m higher than today.
6. Locally, relative sea level will change depending on local and regional land lift or subsidence.
7. At a low, purely technical level of engineering, selection of the most feasible response to sea-level rise presents no new problems.
8. On the higher level of regional and national decision making, however, all kinds of trade-offs will have to be made. This initiates the urgent need for a dedicated decision-support system, combining a large amount of multidisciplinary analysis in a highly orderly and quantitative way, incorporating a conceptual and computational framework and an information system.
9. The ISOS-model developed by Baarse and Rijsberman (1987) shows a useful approach in the required direction.

# References

Baarse, G. and Rijsberman, F. R. (1987) 'Policy analysis', In Wind, H. G. (ed.), *Impact of Sea-level rise On Society (ISOS)*, Report of a project-planning session in Delft, 27–29 August 1986, A. A. Balkema, Rotterdam.

Delft Hydraulics (1988) *Impact of sea-level rise on society, a case study for the Netherlands*, Report H 750 Phase 1 prepared for UNEP and the Government of the Netherlands.

Delft Hydraulics (1989) *Criteria for assessing vulnerability to sea-level rise: a global inventory to high-risk areas*, Report H 838 prepared for UNEP.

Devoy, R. J. N. (ed.) (1987) *Sea Surface Studies. A global view*, Croom Helm, London.

Hulsbergen, C. H. and Schröder, P. C. (eds) (1989) *Republic of Maldives: Implications of sea-level rise*, Report H 926 prepared for the Netherlands Ministry of Economic Affairs, UNDP and the Government of Maldives.

National Research Council (1987) *Responding to changes in sea level; engineering implications*, National Academy Press, Washington, DC.

Plassche, O. van de (ed.) (1986) *Sea-level Research: a manual for the collection and evaluation of data*, Geo Books, Norwich.

Raper, S. C. B., Warrick, R. A. and Wigley, T. M. L. (1988) 'Global sea level rise: past and future', Paper presented at the SCOPE Workshop on Sea Level Rise, held in Bangkok, 9–13 November.

Street-Perrott, A., Beran, M. and Ratcliffe, R. (eds) (1983) *Variations in the Global Water Budget*, Papers presented at the Symposium on Variations in

the Global Water Budget held in Oxford, August 1981, D. Reidel, Dordrecht.
Tooley, M. J. and Shennan, I. (eds) (1987) *Sea-level Changes*, The Institute of British Geographers Special Publication Series 20, Basil Blackwell, Oxford.

# 16 CLIMATE CHANGE AND DESERTIFICATION

Mohamed Abdullah Bin-Afeef
*King Abdulaziz University, Jeddah*

## 16.1 Introduction

According to the UN Environment Programme (UNEP), desertification occurs in 80% of the world's natural pastures in arid areas (3100 million ha) and 60% of the rainfed areas (335 million ha) in addition to 30% of the irrigated land (40 million ha). These areas are inhabited by 20% of the world's population. The desertification problem also affects 85% of the Arab countries in Asia. In Africa the recent drought in the Sudan has affected 3.5 million people and has resulted in the emigration of 1.5 million, in addition to great loss of livestock.

The reasons of desertification are still under investigation. Numerous questions regarding climatic change and its effect on desertification are being asked. Is desertification a phenomenon caused by a cyclic drought period? Is it due to a general climatic trend towards aridity? If there is a climatic change, is it a result of human activities such as man's degradation of the environment?

In this chapter the causes of climatic change are discussed and the possible effects of such a change on desertification addressed. Desertification control is also being analysed from the point of view of greenhouse gases and weather modification.

## 16.2 General circulation of the atmosphere

The study of the general circulation of the atmosphere is concerned with a description and explanation of its behaviour over a long period of time and/or distance. Such a study reveals atmospheric conditions, which are generated by large-scale forces. It is therefore possible to divide these forces into two categories: a very large natural time scale of climatic change and a relatively small one. The first category could include the global differential thermal effects caused by the earth–sun distance, the angle between the earth's rotation axis and the sun's rays, the angular velocity of the earth rotating on its axis and around the sun, and the earth's sea and land distribution. On the other hand, atmospheric air content is known to be a means of climatic change on a small time scale. This is due to its effects on the earth's latitudinal and longitudinal temperature gradient.

Some of the gases that constitute the atmosphere play an important role in its movement and thus general circulation. This is due to their energy absorption and radiation characteristics. Of these gases, water vapour, carbon dioxide and ozone are of the greatest interest for research.

On the question of climatic change due to human activity, scientific concern began with the observation of the increase in global warming which has taken place during this century, estimated to be around 0.7°C. There was a consensus that the main sources of this warming are gases released into the air as a result of human activity, i.e. carbon dioxide ($CO_2$), methane ($CH_4$), nitrogen oxides ($NO_x$), chlorofluorocarbons (CFCs) and ozone ($O_3$).

## 16.3 Impact on climate of a change in atmospheric gas content

In recent years, one of the major concerns in climate studies has been the steady increase in the content of greenhouse gases in the atmosphere and its impact on the atmospheric temperature and climatic changes.

### 16.3.1 Impact of carbon dioxide

Almost half of the global warming is attributed to an increase in $CO_2$ concentration, and measurement of this and the other gases has shown continuing increases in their content over the last decade (Biswas, 1984). Man-induced carbon dioxide has been continuously released into the atmosphere since the beginning of the Industrial Revolution. The monthly values of atmospheric $CO_2$ concentration recorded at Mauna Loa, Hawaii (19°N), and other locations show a steady increase in the annual average (UNEP, 1989 – see also Figure 16.1). There are two distinct trends in these observations. The first is a seasonal variation, in which $CO_2$ concentration decreases in the Northern Hemisphere during the summer. The second is a long-term upwards trend, presumed to be a consequence of human activity. These data show an increase in the annual average to 9% in total $CO_2$ (from 315 ppmv to 345 ppmv between 1958 and 1985). Moreover, it has been estimated that between 50% and 75% of $CO_2$ input into the atmosphere from human activity has remained in circulation with the rest being absorbed by oceans and the biosphere (mainly forests) (Liou, 1980).

$CO_2$ lets through virtually all solar radiation. However, it is a strong absorber in the ˜12–18 μm band of the thermal infrared spectrum. Consequently, an increase in atmospheric $CO_2$ content could result in trapping the thermal infrared radiation rising from the lower atmosphere and could produce a greenhouse effect.

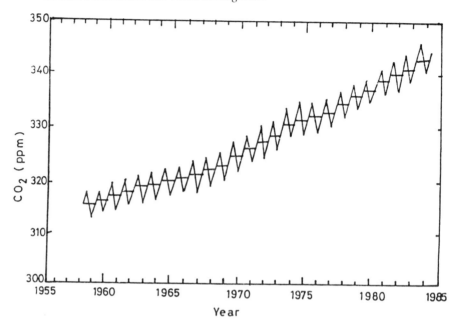

**Figure 16.1**   *Atmospheric concentration of carbon dioxide as measured at Mauna Loa Observatory (UNEP, 1989)*

There have been a number of investigations of the effect of changes in atmospheric $CO_2$ on the atmospheric radiation budget. However, a reliable determination of the changes in atmospheric temperature due to variations in $CO_2$ concentration must take into account not only the radiation aspects of $CO_2$ but also the convective nature of the lower atmosphere. Using a radiative–convective equilibrium model, Manabe and Wetherald (1967) concluded that increases in $CO_2$ concentration have resulted in a warming of the entire lower atmosphere. Based on the assumptions of constant relative humidity and fixed cloudiness and the broadband emissivity values of $H_2O$, $CO_2$ and $O_3$, they found that the changes in mean atmospheric temperature due to $CO_2$ are such that a 10% increase in $CO_2$ concentration would lead to a warming of 0.3°K. A doubling of $CO_2$ concentration from 300 to 600 ppmv produces a 2.36°K increase in the equilibrium temperature ($T_{eq}$) of the earth's surface. Using a three-dimensional general circulation model, Manabe and Wetherald (1975) showed an increase of 2.39°K by doubling the $CO_2$ concentration, a slightly higher value than their earlier results. In addition, they also found that a large cooling occurs in the stratosphere as a result of increased $CO_2$ concentration. Also, tropospheric warming is most pronounced in high altitudes of the lower troposphere, due to the fact that vertical mixing by convection is suppressed in the stable layer of the troposphere in the polar regions. Using an energy balance atmospheric model linked to a one-dimensional upwelling diffusion model of the deep ocean, Michael *et al.*

(1981) showed that in the year 2020, the $CO_2$ concentration is calculated to be 501 ppmv. $T_{eq}$ is 2.22°K and the climatological $T_{air}$ is 1.32°K. Such a temperature lag could delay perception of a climatic change. More recently, atmospheric general circulation models are being used to predict global trends, taking into consideration thermal feedback effects as the cloud cover changes (Chess *et al.*, 1989).

The climate change within the next century could be as large as that between the last Ice Age and the present. An estimated increase of 1.5–4.5°C by the year 2050 in global average temperature is considered very likely. Analysis of historical records of surface measurements suggests that the temperature of the planet has increased by about 0.5–0.7°C during this century (Biswas, 1984).

### 16.3.2 Impact of ozone depletion

The existence of the earth's ozone layer and its importance to life on the planet have been known for many years. It is also believed that all life has evolved and adapted over millions of years to a specific level of ozone in the atmosphere. Only recently has it been realized that some of man's activities could lead to stratospheric pollution and to a significant reduction in the ozone layout over the next few decades.

A signficant characteristic of ozone is its absorption of ultraviolet solar radiation. This has a maximum intensity at a height of about 25 km, the area of greatest ozone concentration. This absorption provides the heat source which maintains a stable stratosphere (the barrier to the turbulent weather system in the troposphere) and is thus one of the ultimate controls on global climate. More importantly, this absorption process shields the earth's surface from ultraviolet radiation, which could have detrimental effects on many forms of life. Since this chapter is mainly concerned with climatic change and desertification, emphasis here will be on the impact of stratospheric ozone depletion on climate and weather.

Interest in the subject began with the discovery of a hole in the stratospheric ozone layer over Antarctica (Farman *et al.*, 1985). Many stations have now been established to measure global ozone concentrations, including the Arctic and Antarctic regions. The following is a review of some research activities on the possible effects on climate of ozone reduction.

Research into ozone fluctuations and their effect on climate has been carried out for many years (e.g. Dobson *et al.*, 1946; Reed, 1950). More detailed analysis of ozone variability has become possible with the advent of satellite measurements. For example, Schoeberl and Krueger (1983) successfully modelled observed ozone changes in the Southern Hemisphere and suggested that, in general, atmospheric disturbances which are vertically trapped near the tropopause and have little vertical phase shift will tend to produce the strongest correlation with total ozone. Recently, detailed satellite observations of ozone have permitted, for the first time, almost spatially complete estimates of the climatology of the total ozone (Bowman and Krueger, 1985). Such data give a broader basis for research work into ozone and climate. For example, Rodgers *et al.* (1986) used total ozone to monitor the interaction between tropical storms and their environment. Sechrist *et al.* (1986) identified secondary circulation features associated with a mid-latitude jet stream. Schubert *et al.* (1988) studied the relation between total ozone and tropopause pressure for the four years 1979–1982. Kiehl *et al.* (1988) used the NCAR climate model to study radiative and dynamic response to impose changes to the model ozone distribution for the month of January. It is deduced that for a 75% reduction in ozone in the Northern Hemisphere, the jet stream is severely reduced. Much work still needs to be done to clarify the change in ozone content and its relation to weather and climate.

## 16.4 Desertification

Desertification arises from the interaction between a dry land environment and man's use of it. An understanding of the controlling mechanisms of climates in dry areas helps us to appreciate climatic factors in desertification. The result of the global circulation pattern is that the subtropics (15–30° North and South) are regions of subsiding air which inhibit rain formation. However, dry climates are also extended to other latitudes as a result of additional factors such as distance from rain-supplying oceans, high-pressure zones of large continental areas linked with monsoon systems and the presence of mountain barriers. The world desertification map (Figure 16.2) shows the distribution of dry climates. It is believed that the boundaries of dry land were not fixed (UN, 1977), and that many areas which are now dry maintained vegetation cover in the distant past.

Climatic changes have been shown to be part of global shifts of the climate belts related to variations in the earth's atmospheric circulation and energy budget. The present dry climates in which desertification is most widespread are defined by means of the Budyko–Lettau dryness ratio, defined as the ratio of the annual net radiation at the surface to the energy required to evaporate a year's normal rainfall. Figure 16.3 shows the distribution of this ratio throughout the world. The areas subject to desertification lie between values of 2 and 7.

As mentioned above, aridity arises from persistent atmospheric subsidence. It may also be caused by the absence of humid airstreams and of rain-inducing disturbances in the atmosphere. Clear skies and low humidities in most regions give dry climates very high solar radiation incomes, which lead to high soil temperatures.

The subtropical high-pressure belts migrate towards the poles in summer and the equator in winter. This gives a threefold structure to the arid zone (Figure 16.4): a Mediterranean fringe with rains only in winter, a

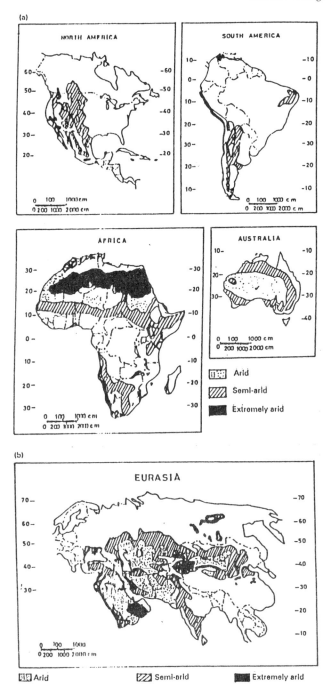

**Figure 16.2** *The distribution of extremely arid, arid and semi-arid lands in (a) North and South America, Africa and Australia and (b) Eurasia. (Reproduced with the permission of the AAAS and of H. E. Dregne, editor of* Arid Lands in Transition, *AAAS, Washington, DC, 1980)*

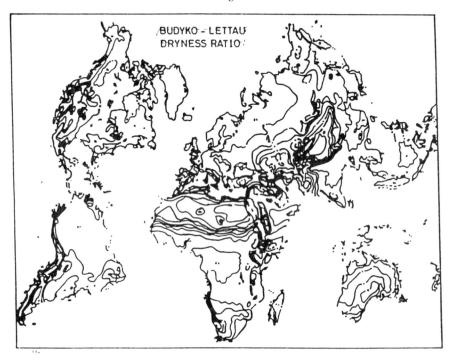

**Figure 16.3**   *The Budyko–Lettau dryness ratio over the earth's land areas (reduced and simplified from the original computations by Henning). The ratio shows the number of times the mean annual net radiation income could evaporate the local mean annual rainfall. All detail has been omitted in humid areas (i.e. where the ratio is less than unity)*

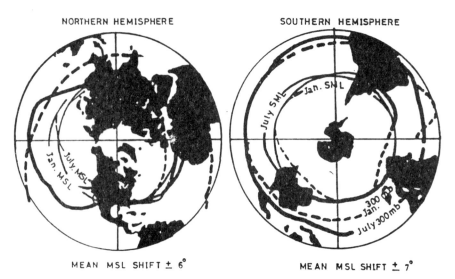

**Figure 16.4**   *January to July shifts of the axis of the subtropical high-pressure belt at 300 mb (approximately 9.1 m above sea level) and at sea level. Note that the belt moves towards the poles in the summer and the equator in the winter*

desert core with little or no rain (20–30° latitude) and a tropical fringe with rain mainly in the high-sun season.

Recent climatic variations such as the Sahelian drought are natural in origin and are not unprecedented. Statistical analysis of rainfall shows a distinct tendency for abnormal wetness or drought to persist from one year to the next. This suggests that feedback mechanisms may be at work whereby drought feeds drought and rain feeds rain. General circulation models aim to simulate the statistics of the earth's dynamic climate. When dry surfaces are degraded by overstocking or harmful cultivation, albedo increases. Several models suggest that this should also increase subsidence and hence aridity.

## 16.5 Impact of climate change on desertification

In spite of the uncertainty which surrounds the magnitude, timing and spatial characteristics of climate change and the impacts these changes may have on desertification, it is not too early to consider their implications on desertification. The world has experienced an abnormally long series of dry years in the Sahel over the period 1970–1984. This gives us an indication of what might happen in arid and semi-arid areas following climate change-induced shifts in the agro-climatic zones. Even in the absence of any climatic change, man's misuse of his environment can lead to gradual deterioration of land and consequently more extended desertification.

At present, climate change modelling and impact analysis reveal that the serious implications of climate change on desertification are not at the global level, where the net effect is likely to be positive, but at the national and local levels in many low-income developing countries. General circulation models are unable to predict regional or local effects with an acceptable degree of certainty. In addition, the directions of possible change in the climate in the middle latitudes are much more uncertain than those in the north (FAO, 1989).

One of the accepted predictions of climate change is that the greenhouse effect will not be uniform everywhere. Warming at the poles is likely to be 200–300% of the global average, while it may be only 50–100% near the tropics (Biswas, 1988). It is also pointed out that as the natural thermal gradients on the planet's surface are reduced, global patterns of winds and ocean currents as well as the timing and distribution of rainfall will change. Ocean currents that now moderate the climates of certain high-latitude countries may shift, causing cooling effects in some countries even as the rest of the world experiences warming. Specifically, the climatic change will effect desertification in two ways:

1. *Shifts in the agro-climatic zones*: Broadly, desertification is expected to increase in hotter, drier interiors. This is because of decreased rainfall and increased evapotranspiration as a result of the expected shift in the subsidence belt of atmospheric general circulation. On the other hand,

the boundaries of existing cold continental deserts will sustain more agricultural activity mainly as a result of the expected substantial warming at high latitudes discussed above.

2. *Land loss through sea-level rise:* Since some of the most productive areas are low-lying coastal plains and estuaries with fertile alluvial soils, sea-level rise could be a major threat. In Asia, for example, a high proportion of rice production is in low-lying coastal areas, mainly former swamps and marshes. Losses could occur through submergence, increased incidental damage from cyclones, longer or deep fresh-water flooding in some inland parts and increased salt-water intrusions in coastal aquifers used for irrigation. This may be a result of sea-level rise; reduced recharge rates of aquifers due to declining precipitation; or of failure to adjust groundwater extraction rates to balance recharge rates. Such events reduce the area available for cultivation.

## 16.6 Desertification control

*16.6.1 Response strategies to greenhouse gas emission*

There is general agreement that the increase in global warming which has taken place this century is consistent with predictions on the emission of greenhouse gases due to human activity. At this rate, a further warming of the earth to the extent of 1.5°–4.5°C by the middle of the next century is very likely. Thus greater cooperation among all nations will be needed to successfully implement measures to reduce the level of emissions of greenhouse gases. These measures must be in two areas:

1. *Control the sources of greenhouse gases:* The important atmospheric pollutants and greenhouse gases are carbon dioxide ($CO_2$), chlorofluorocarbons (CFCs), methane ($CH_4$), tropospheric ozone ($O_3$), and nitrogen oxides ($NO_x$). The sources of these emissions are clearly attributed to three main activities which should be controlled (IPCC, 1989), i.e. agriculture, industry and fossil-fuel combustion.

   In agriculture, the sources of emission of greenhouse gases are fertilizers, pesticides and animal wastes. Improved technology and management could lead to substantial reductions of these gases in the atmosphere.

   Industry's contribution to the greenhouse gases is the production of halocarbons and creation of landfill sites. The halocarbon industry is characterized by the existence of many different pollutants. In some cases the environment will not be able to stand the ecological load imposed by halocarbons at any emission level and only a total ban would be appropriate. Methane emissions could be limited by improving the management of landfill sites and reducing the waste stream, and by supporting the production and sales of methane gas from these sites.

The emission of greenhouse gases from fossil-fuel combustion is inevitable. The use of fossil fuel in human activity is a necessity, at least in the near future. Other alternatives are still not economically feasible. Clearly, the problem is attributed to inefficient use of fossil fuel. Measures should be taken towards improving combustion efficiency and reducing emission of the residual gases by proper filtration.

2. *Forestry as a sink for greenhouse gases:* It is generally agreed that higher carbon dioxide levels in the atmosphere will increase the rate of photosynthesis of $C_3$ plants and lead to a reduction of water loss from both $C_3$ and $C_4$ plants (Liss and Crane, 1984). Cure and Acock (1986) state that the overall net $CO_2$ exchange rate of crops increased by 52% on first exposure to doubled $CO_2$ concentration, but was only 29% higher after the plants had acclimatized to the new concentration. Thus, forests are major greenhouse sinks all over the world, and reforestation should be encouraged and deforestation restricted.

## 16.6.2 *Response strategies to sea-level rise*

In areas where sea-level rise could be a major threat to the most productive low-lying coastal plains, it may be economic to consider coastal protection, stormsurge and tidal barriers and other forms of flood control. Such measures, however, could be very expensive. Therefore the building of higher coastal defences would only be feasible if additional international financial assistance were forthcoming (FAO, 1989).

## 16.6.3 *Modification of atmospheric circulation in the boundary layer*

As the weather conditions in the boundary layer are affected by the earth's surface, any alteration of the physical nature of this surface will be reflected in the solar energy budget and consequently in the atmospheric circulation in the boundary layer. Three modification schemes were suggested for changing the physical characteristics of the atmospheric boundary layer:

1. *Rain enhancement:* Cloud seeding and the study of cloud physics for the purpose of rain enhancement have been widely discussed (Olsen and Woodley, 1975; Sevruk, 1982, 1985; Rindsberger, 1985; Robichand and Austin, 1988). It is generally accepted that the problem is one of increasing the clouds' efficiency in processing vapour in the atmosphere. Cloud seeding is used to increase precipitation from clouds that were already producing or were about to produce rain. Rain enhancement has been carried out in the United States, the USSR, Brazil, Canada, France, Japan, Australia and Zaire with varying degrees of success (Changnon, 1975). The question is: 'What is the increase in rainfall that can be expected from cloud seeding?' This points to the need for additional research on the subject.

There has been a proposal for water transfer from the water-rich tropical parts of Africa to water-poor regions such as the Sahel (Falken-mark and Lindh, 1974). The plan involved redistribution of prevailing rainfall by displacement of the tropical fronts, resulting in more rain in Africa's interior.

2. *Carbon dust:* The carbon dust proposal is similar in principle to the asphalt concept. The major difference is that the artificial heat source would be placed directly into the atmosphere rather than on the earth's surface (Gray *et al.*, 1974). The carbon in the air acts as an artificial heat source to increase evaporation if released over the water surface. The advantage of carbon dust is that the area of its impact is much greater (10 000–100 000 km$^2$) than that for asphalt (100 km$^2$) (Gray, 1976). The technique has not yet been tested. Its disadvantage is the long-term side-effect of carbon dust on the environment.

3. *Asphalt island concept:* There is usually water vapour in the atmosphere over arid or semi-arid lands, but the mechanism which transforms this water vapour into rainfall is lacking in these regions. It was suggested (Black, 1970) that significant additional rainfall can be produced in some areas by spraying a large area near a water body with a thick coating of asphalt. The surface albedo (reflectivity) will be much lower than that of the surrounding sand, soil, vegetation or water. The temperature difference between coated and uncoated surfaces would lead to a convective motion of air above the asphalt surface. This would promote cloud formation and precipitation if sufficient moisture were present in the atmosphere, and would produce an effect similar to that caused by mountain ranges but with a smaller magnitude. This suggestion is based on theoretical considerations, and has yet to be tested in practice (Gray *et al.*, 1974).

4. *Other means:* Another scheme, taken seriously in the 1950s, suggested the transportation of icebergs from the Antarctic to the western coastal waters of South Africa, which would cause an increase in precipitation over the coastal belt and further inland (Taljaard, 1976). A recent proposal to tow icebergs from the Antarctic to the Arabian coast has also been considered by government officials in Saudi Arabia (*Christian Science Monitor*, 1975).

# Acknowledgements

The author would like to thank Dr Ayman Abdullah and Dr Mostafa Ibrahim of the Department of Meteorology, Faculty of Meteorology, Environment and Arid Land Agriculture, King Abdulaziz University, Jeddah, for their assistance, valued advice and constructive comments in the preparation of this chapter, without which it would not have been completed. The author is also indebted to Mr Mohamed Nasr Allam and Dr Mohamed Jameel Abdulrazzak for their support and encouragement.

# References

Biswas, A. K. (1984) *Climate and Development*, Cassell Tycooly, London, p. 141.

Block, J. F. (1970) *Asphalt Island Concept of Weather Modification*, Linden, NJ, Exxon memorandum, 4 June.

Bowman, K. P., and Krueger, A. J. (1985) 'A global climatology of total ozone from NIMBUS-7 total ozone mapping spectrometer', *J. Geophys. Res.*, **90**, 7967–7976.

Chess, R. D. *et al.* (1989) 'Interpretation of cloud–climate feedback as produced by 14 atmospheric general circulation models' *Science*, **245**, 513–516.

Cure, J. D. and Acock, B. (1986) 'Crop responses to carbon dioxide doubling: A literature survey', *Agricultural and Forest Meteorology*, **38**, 127–145.

Dobson, G. M. B., Brewer, A. W. and Cwilong, B. M. (1946) 'Meteorology of the lower stratosphere', *Proc. Roy. Soc. London*, **A185**, 144–175.

Dunkerton, T. J. (1988) 'Body force circulation and the Antarctic ozone minimum', *J. Atmos. Sci.*, **45**, 427–438.

FAO (1989) 'Implications of climate change for food security', Theme paper for FPCC Working Group III sub-group on Resource Use and Management.

Farman, J. C., Gardiner, B. G. and Shauklin, J. D. (1985) 'Large losses of total ozone in Antarctica reveal seasonal $ClO_x/NO_x$ interaction', *Nature*, **315**, 207–210.

Gray, W. *et al.* (1974) 'Weather modification by carbon dust absorption of solar energy', *Proceedings of the 4th Conference on Weather Modification*, Fort Lauderdale, Florida, 18–21 November, p. 195.

Gray, W. M., Frank, W. M., Covin, M. L. and Stokes, C. A. (1976) 'Weather modification by carbon dust absorption of solar energy', *J. Appl. Met.*, **15**(4); 355–386.

IPCC (1989) 'Economic measures as a response to climate change'. Theme paper for IPCC Response Strategies Working Group.

Kiehl, J. T. and Boville, B. A. (1988) 'The radiative–dynamical response of a stratospheric–tropospheric general circulation model to changes in ozone', *J. Atmos. Sci.* **45**, 1798–1817.

Liou, Kuo-Nan (1980) *An Introduction to Atmospheric Radiation*, Academic Press, New York.

Liss, P. S. and Crane, A. J. (1984) *Man-made Carbon Dioxide Climatic Change*, Geo Books, Norwich.

Manabe, S. and Wetherald, R. T. (1967) 'Thermal equilibrium of the atmosphere with a convective adjustment', *Journal of Atmospheric Science*, No. 21, 361–385.

Manabe, S. and Wetherald, R. T. (1975) 'The effects of doubling the $CO_2$ concentration on the climate of a general circulation model', *Journal of Atmospheric Science*, No. 32, 3–15.

Michael, P., Hoffert, M., Tobias, M. and Tichler, J. (1981) 'Transient climate response to changing Carbon Dioxide concentration'. *Climatic Change*, **3**, 137–153.

Reed, R. J. (1950) 'The role of vertical motions in ozone–weather relationships', *J. Meteor.*, **7**, 263–267.

Rodgers, E., Stout, J. and Steranka, J. (1986) 'Upper-tropospheric lower-stratospheric dynamics associated with tropical cyclones as inferred from total ozone measurements', *Second Conf. Satellite Meteorology/ Remote Sensing and Applications*, Williamsburg, Amer. Meteor. Soc.

Schoeberl, M. H. and Krueger, A. J. (1983) 'Medium scale disturbances in total ozone during southern hemisphere summer', *Bull. Amer. Meteor. Soc.*, **64**, 1358–1365.

Sechrist, F. S., Petersen, R. A. Brill, K. F., Kruger, A. J. and Vecellini, L. W. (1986) 'Ozone, jet streams and severe weather', *Second Conf. on Satellite Meteorology/Remote Sensing and Applications*, Williamsburg, Amer. Meteor. Soc.

Snieder, R. K. and Stephen, S. B. (1988) 'The flywheel effect in the middle atmosphere', *J. Atmos. Sci.*, **45**, 3996–4004.

Taljaard, J. J. (1976) Republic of South Africa Weather Bureau, Pretoria, 27 January.

United Nations (1977) *Desertification: its Causes and Consequences*, Conference held in Nairobi, Kenya, 29 August to 9 September.

United Nations Environmental Programme (1989) *The State of the World Environment*, UNEP/GC.15/7/Add.2.

*Part 3*

# Planning and Management

# 17 RECENT AND FUTURE PRECIPITATION CHANGES OVER THE NILE BASIN

M. Hulme
*University of East Anglia, Norwich, UK*

## 17.1 Introduction

The Nile Basin encompasses nine countries in north-east Africa and has a surface area of just over 4 million $km^2$. The river itself is 6640 km from source to mouth, the longest in the world. The population of the Nile Basin is about 140 million, of whom at least 50% are heavily dependent on the Nile waters for their economic and domestic existence. The reliability of Nile discharge is fundamental therefore for the well-being of north-east Africa, and especially of Egypt and the Sudan, the two major downstream nations in the Nile Basin.

Nile discharge has fluctuated both historically and prehistorically. Well-established periods of substantially lower flows than present occurred during the last glaciation around 20 000 BP (Williams and Faure, 1980), and within the Holocene smaller fluctuations occurred, resulting both in increased (e.g. 6000–4000 BP; Wickens, 1975) and decreased discharge (e.g. AD 1180–1350; Hassan, 1981). In the twentieth century substantial interannual variability in Nile discharge has occurred with a maximum annual yield of 120 $km^3$ in 1916 and a minimum of only 42 $km^3$ in 1984. Interdecadal variations in the present century have also been substantial, with mean annual discharge between 1900 and 1959 reaching 84 $km^3$ compared to a mean of only 72 $km^3$ between 1977 and 1987.

Such fluctuations in discharge reflect a variety of both natural and anthropogenic causes operating on different time scales. Natural factors include changes in precipitation regimes (particularly over the headwaters of the Blue and/or White Niles), in evaporation, and in vegetation in the catchments which affect runoff. Such changes, however, may also be anthropogenic in origin, with precipitation being affected by, for example, human-induced land-cover change (Nicholson, 1988), evaporation by the creation of artificial water bodies, and vegetation by the extraction of woodfuel (Ahlcrona 1988). Additional anthropogenic factors affecting the total Nile discharge include the extraction of water for domestic, agricultural, industrial, or power-generation purposes.

This chapter focuses on one of these controls on Nile discharge which may be both natural and anthropogenic in origin, namely, variations in the precipitation input over the Nile Basin. We first analyse recent precipitation variations in the twentieth century using an extensive monthly precipitation data set containing just under 200 station records. Such recent precipitation changes are thought to have resulted largely from natural variations in the climate system, although with increasing anthropogenic interventions over the last few decades. Thus present explanations for variations in Nile Basin precipitation focus on changes in global sea surface temperatures (SSTs), land-cover changes, and variations in the behaviour of the Tropical Easterly Jet (TEJ).

Second, we consider the possible future precipitation inputs over the Nile Basin. Here, the anthropogenic forcing of global climate due to increasing atmospheric concentrations of greenhouse gases is likely to dominate climate changes in the twenty-first century. Although the poor resolution and the still-incomplete physical representation of climate in the current generation of atmospheric general circulation models (GCMs) precludes the development of accurate regional climate *predictions*, we present a regional *scenario* based on climate experiments using the five leading GCMs. By compositing the results of these five model experiments, we can construct a single model temperature and precipitation scenario. This may be regarded as the current best-guess of a future equilibrium climate for the Nile Basin following a doubling of greenhouse gas concentrations.

## 17.2  Recent precipitation changes

The monthly precipitation data used in this chapter are from the Global Precipitation Dataset (GPD) held in the Climatic Research Unit at the University of East Anglia. This GPD has been extended and updated from that used by Bradley *et al.* (1987) and Diaz *et al.* (1989). Data are held for just under 200 stations in the Nile Basin with records commencing in the 1880s and continuing through 1989 (Figure 17.1).

Precipitation series for three selected regions are constructed using a procedure outlined in Hulme (1989) and precipitation data are transformed to probability values using a root-normal distribution. Within each respective region the data are then averaged to produce a single probability value. This is converted back to mms using calculated region-average distribution parameters. This procedure produces precipitation series which are relatively insensitive both to missing data and to changes in station networks over time. The three regions selected are 5° latitude by 5° longitude boxes and represent the upper White Nile catchment (referred to henceforth as 'Uganda'), the upper Blue Nile catchment ('Ethiopia'), and the middle Nile Basin including the confluence of the White and Blue Niles at Khartoum ('central Sudan') (see Figure 17.1). The Blue Nile contributes between 70%

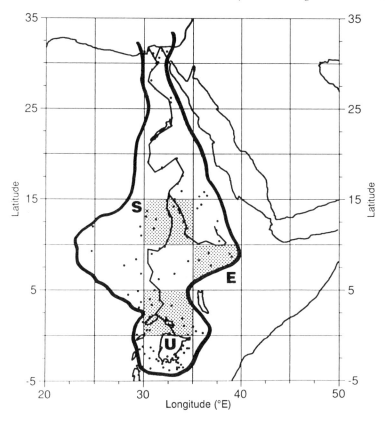

**Figure 17.1** *The Nile Basin, precipitation stations in the GPD, and the three 5° by 5° regions used in the analysis (S = central Sudan; E = Ethiopia; U = Uganda). Only those stations within the Nile catchment (shown by the heavy line) are included. The temperature curve in Figure 17.5(b) is for the whole area covered by this figure*

and 80% to overall Nile discharge, and the White Nile between 20% and 30%. The contribution of precipitation over central Sudan to Nile discharge is small.

The twentieth-century precipitation time series for these three regions is shown in Figure 17.2. Owing to differences in seasonality between the regional precipitation regimes, annual precipitation is shown for Uganda, whereas seasonal precipitation from June to August (JJA) is shown for the other two regions. JJA seasonal precipitation represents 54% and 70% of annual precipitation in Ethiopia and central Sudan, respectively.

Table 17.1 shows the direction of precipitation change over the twentieth century calculated by a simple linear regression. All three regions show a decline in precipitation totals, although this decline is minimal for Uganda. In relative terms, the decline is greatest in central Sudan, where a 9% reduction in seasonal precipitation has taken place over the twentieth

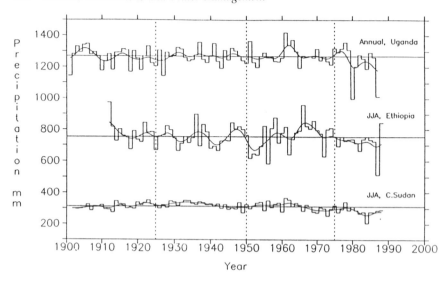

**Figure 17.2**   *Precipitation time series for the three regions within the Nile Basin. Horizontal lines indicate median precipitation over the full record*

**Table 17.1**   Regression analysis for three regional precipitation time series

| Region | Series | Length | Median precip. (mm) | Slope (mm/yr) | Percentage change over length of series |
|---|---|---|---|---|---|
| Uganda | Annual | 1901–1988 | 1269 | −0.27 | −1.8 |
| Ethiopia | Seasonal | 1912–1988 | 754 | −0.48 | −4.9 |
| Central Sudan | Seasonal | 1902–1988 | 312 | −0.33 | −9.2 |

century. In all three regions, however, most of this reduction has occurred within the last two decades, with substantial negative anomalies being recorded since 1965, 1975 and 1980, respectively, for central Sudan, Ethiopia and Uganda (Figure 17.2). The precipitation peak in the early 1960s in Uganda and the mid-1960s in Ethiopia contributed to the high levels of Lake Aswan over the first decade after its completion such that its peak level of 177.41 m above mean sea level was reached in November 1978. The subsequent decade saw falling levels in lake Aswan, which were only reversed in 1988 following a high Blue Nile discharge (Sutcliffe *et al.*, 1989). The very low precipitation input in 1987 in both Uganda and Ethiopia and the high 1988 input in Ethiopia is evident from Figure 17.2.

One of the intriguing features of the Nile system is the contrasting nature of the precipitation regimes which supply the headwaters of the two branches of the Nile. The White Nile is fed from equatorial rains which possess a weak bimodal seasonal distribution, while the Blue Nile is supplied from a seasonally arid regime over the Ethiopian highlands. The meteorology controlling these two regimes is quite distinct (e.g. Griffiths, 1972) and the correlation between the annual precipitation input to these two catchments is therefore not necessarily good.

Figure 17.3 shows the association between annual precipitation over Ethiopia and Uganda for the period 1914–1988. There is effectively zero correlation between these two regimes, implying that there is neither a synchronous (positive correlation) nor asynchronous (negative correlation) tendency in the interannual fluctuations of the two precipitation regimes which contribute to main Nile discharge. In certain years high precipitation input to one catchment may be offset by low input to the other (e.g. 1980), while in other years both catchments may be poorly supplied (e.g. 1987), or well supplied (e.g. 1963). The high Nile discharges of the 1960s are reflected in the high precipitation inputs in both catchments and contrast with the generally poorer precipitation performance in the 1980s (Figure 17.3).

Although precipitation over central Sudan contributes little to overall Nile discharge the negative precipitation anomalies of the last two decades have been most marked in this region. Central Sudan is part of the Sahel region of Africa, and precipitation here is particularly sensitive to a range of factors which control the extent and magnitude of the African monsoon (Farmer, 1989). It is therefore useful to compare the precipitation changes

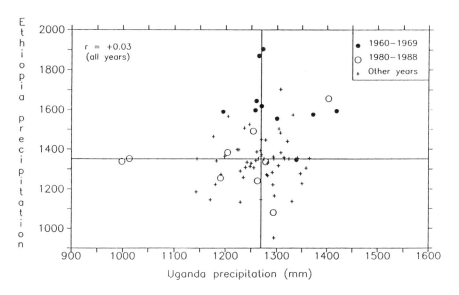

**Figure 17.3** *Annual precipitation for Uganda plotted against that for Ethiopia (1914–1988). The 1960s and 1980s are separately identified*

of the present century in this region with those estimated to have occurred during the last 20 000 years. Figure 17.4 compares the position of the 400 mm annual isohyet in central Sudan for a number of years, periods, and centuries. Taking the driest and wettest 30-year periods in the twentieth century produces a latitudinal shift in precipitation belts in central Sudan of between 50 km and 75 km. This movement increases to between 200 km and 300 km when individual years (1929 and 1984) are considered. By comparison, latitudinal shifts in the Holocene have been up to 400 km in extent, although these are estimated century-mean positions (Figure 17.4). The sensitivity of 30-year precipitation regimes in central Sudan to forcing mechanisms in the present century therefore represents up to 25% of that experienced by 100-year precipitation regimes during the Holocene period in which substantial changes in Nile discharge are known to have occurred (Adamson et al., 1980).

## 17.3 Future precipitation changes

At least four different approaches to estimating future precipitation levels over the Nile Basin can be pursued. First, short-term seasonal forecasts of precipitation can be made using global fields of sea surface temperature (SST) anomalies (Parker et al., 1988). This method is being further developed, and currently is more relevant for the Sahel zone of central Sudan rather than for the Ethiopian or Ugandan catchments. Future extensions of this method may, however, include such regions (Ward, N. M., pers. comm.). Second, analysis of the instrumental time series of precipitation may generate a statistical basis for forecasting future precipitation changes. The anecdotal seven-year flood/drought cycle of the Nile or the proposed 18.5-year lunar cycle in Nile discharge (Currie, 1987) are examples of such an approach. However, evidence for robust statistical cycles in the precipitation record over either Ethiopia or Uganda is weak. Furthermore, such statistical forecasts assume a past and future constancy in the boundary conditions which control the climate of the Nile Basin. Such an assumption is increasingly unsustainable. Third, historical analogues may be constructed from which future precipitation scenarios for the Nile Basin may be produced. For example, the proposed relationship between cold periods in Europe and low Nile flows (e.g. Hassan, 1981), or the reconstructions of African precipitation during the Holocene (COHMAP, 1988) or for the latter end of the Pleistocene (e.g. CLIMAP, 1976) are examples of such an approach. A fourth method of estimating future precipitation over the Nile Basin involves the creation of climate scenarios from GCM experiments. In this section we present some results using this latter approach.

There is a broad consensus that the major factor determining global climate in the twenty-first century will be the concentration of greenhouse gases in the atmosphere. As this concentration increases, the energy flux within the atmosphere will be altered so as to lead to a warming of the

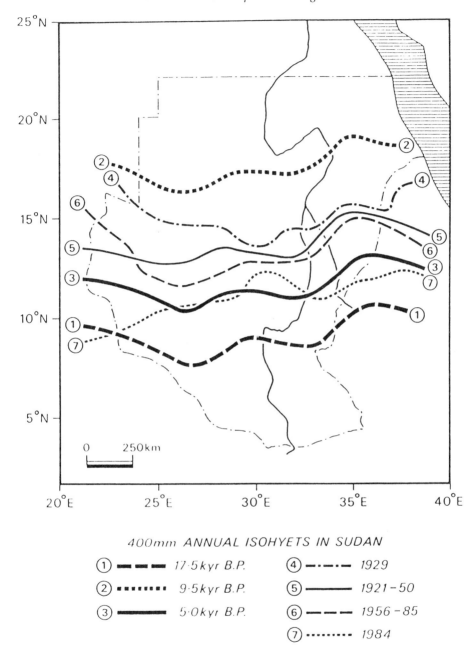

**Figure 17.4** *Position of the 400 mm annual isohyet for various twentieth-century years and periods and for three Holocene centuries. Holocene estimates are from Wickens (1975, 1982)*

near-surface air temperature and a cooling of the stratosphere. Since greenhouse gases have been increasing steadily over the last 100 years, some evidence for global warming should already be evident in the instrumental temperature record. Figure 17.5(a) shows the time series of global annual surface air temperature from 1861 to 1988 averaged for both land and marine areas (Jones *et al.*, 1988). An overall global warming of about 0.5°C is evident. The regional dimensions of this warming need not be uniform, however, and Figure 17.5(b) shows the equivalent time series for 1871–1988 for the Nile Basin (defined as in Figure 17.1). An overall warming of again about 0.5°C is evident, although most of this warming occurred in the early decades of this century, with little change in Nile Basin temperatures since the 1950s.

Estimating the future rate and pattern of global warming and its consequences for other climatic parameters such as precipitation is best accomplished at present using GCMs. These are employed in experiments which generate two global climates, one representing climate under current (1 × $CO_2$) greenhouse gas concentrations, and the second representing climate under doubled (2 × $CO_2$) concentrations of greenhouse gases. The difference between the 2 × $CO_2$ and 1 × $CO_2$ climates is the estimated greenhouse climate signal. Different GCMs, however, use different spatial resolutions and different physical parameterizations by which the atmosphere is simulated. Their resulting climates are therefore not similar, and no single GCM can be regarded as necessarily better than another. A method to utilize the simulated greenhouse climates of different GCMs and produce a composite model climate scenario has been developed by Wigley *et al.* (1990). The details of this method are not discussed here, but the following composite GCM scenario for the Nile Basin is derived from the five leading GCM experiments, namely, those performed at the Goddard Institute for Space Studies, the Geophysical Fluid Dynamics Laboratory, Oregon State University, the National Centre for Atmospheric Research and the UK Meteorological Office (see Wigley *et al.*, 1990). The scenario presents the equilibrium change in climate which could result from a doubling of $CO_2$-equivalent (i.e. from 300 ppmv to 600 ppmv). The date by which such a doubling would occur is uncertain, but probably lies between the years 2030 and 2070 (Bolin *et al.*, 1986).

Figure 17.6 shows the change in mean seasonal air temperature over the Nile Basin for the winter (DJF) and summer (JJA) seasons. Winter temperatures increase by between 3° and 4°C over the Nile Basin, summer temperatures by slightly less at between 2° and 3°C. For summer, the temperature increase is greatest over the lower Nile Basin. Associated with such temperature increases will be a rise in potential evapotranspiration ($ET_p$). Changes in actual evapotranspiration ($ET_a$) will depend on moisture availability, but will undoubtedly lead to greater evaporative water loss from the Nile Basin and consequently reduced Nile discharge. Will this increased evaporative water loss be offset by increased precipitation?

Figure 17.7 shows the change in seasonal precipitation rate (in mm/day) for the winter (DJF) and summer (JJA) seasons over the whole Nile Basin

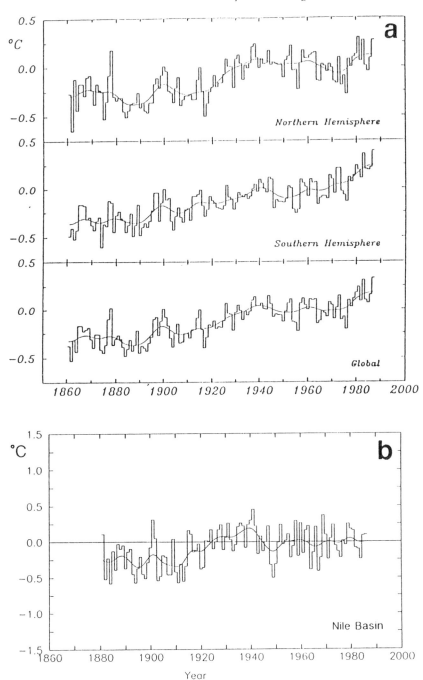

**Figure 17.5**   *Annual time series of surface-air temperature based on land and marine data. (a) Global and hemispheric series (from Jones et al., 1988); (b) Nile Basin (as defined in Figure 17.1)*

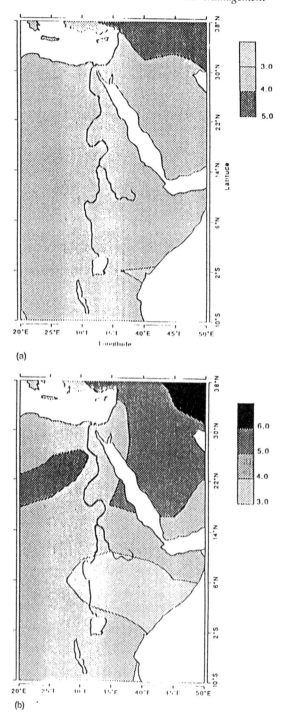

(a)

(b)

**Figure 17.6**   *Model average of the* $2 \times CO_2 - 1 \times CO_2$ *change in surface air temperature for the Nile Basin. (a) DJF; (b) JJA. Contour interval is 1°C*

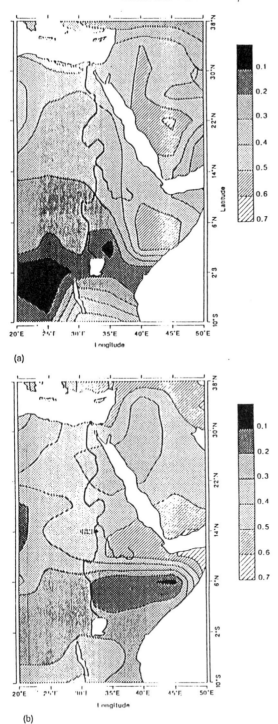

(a)

(b)

**Figure 17.7**   *Probability of a decrease in precipitation following a doubling of* $CO_2$-*equivalent, derived from five GCM experiments. (a) DJF; (b) JJA. Contour interval is 0.1*

and Table 17.2 summarizes the projected changes for the three regions analysed previously. Precipitation changes are presented in three ways in Table 17.2: as seasonal change in total precipitation yield, as a percentage change in existing yield, and as a probability that precipitation will *decrease* by some unspecified amount. The last value is calculated from the assumption that the five absolute precipitation changes derived from the GCMs are sampled from a normal distribution, thus enabling a theoretical probability for a given precipitation change to be estimated. This probability estimate in effect gives an indication of the confidence we have in the direction of the projected precipitation change derived from the five GCM experiments. Where the probability is high (low) the models generally agree that precipitation will decrease (increase).

**Table 17.2**   Seasonal changes in precipitation (absolute, relative and probability) for a doubling of $CO_2$-equivalent derived from the composite GCM climate scenario

|  | DJF | | | JJA | | |
|---|---|---|---|---|---|---|
|  | *Absolute change (mm)* | *Change (%)* | p *(precip. decrease)* | *Absolute change (mm)* | *Change (%)* | p *(precip. decrease)* |
| Uganda | +36 | (+28.1) | 0.1 | +63 | (+18.2) | 0.15 |
| Ethiopia | 0 | (–) | 0.5 | −9 | (−1.2) | 0.55 |
| Central Sudan | +9 | ($\infty$) | 0.35 | 0 | (–) | 0.5 |

Figure 17.7 and Table 17.2 indicate contrasting projections of seasonal precipitation change between equatorial Africa (Uganda) and seasonally arid Africa (Ethiopia and central Sudan). For Ethiopia and central Sudan no clear greenhouse precipitation signal emerges from the five experiments. Absolute and relative precipitation change is close to zero and the probability of a precipitation decrease is close to 0.5. There is a hint that summer precipitation over the Blue Nile catchment may decrease slightly and winter precipitation over central Sudan may increase slightly. The latter result shows the danger of treating GCM scenarios as realistic predictions of climate change. Contemporary winter precipitation over central Sudan is zero, thus projecting an increase of 0.1 mm/day would require a major change in the seasonality of the monsoon circulation and seems an unlikely possibility. The reason for this, of course, is that the GCM experiments which simulate present climate do so inadequately and generate, for example, winter precipitation in central Sudan which in reality is non-existent.

The model signal over Uganda is somewhat clearer. The projected probability of a precipitation decrease is only 0.1 (DJF) and 0.15 (JJA) with estimated absolute precipitation increases of 36 mm (+28%) and 63 mm

(+18%), respectively. This contrasting result between the White and Blue Nile catchments is not surprising, since the meteorology of the two catchments differs substantially, and the instrumental precipitation record over the twentieth century demonstrates zero correlation between Uganda and Ethiopia precipitation (Figure 17.3).

## 17.4 Conclusions

In the twentieth century the White and Blue Nile catchments have witnessed different trends in annual precipitation yield. The White Nile has experienced little long-term trend in annual yield (an overall decline of less than 2%), although notable short-term episodes of unusual precipitation occurred in the early 1960s (high) and the 1980s (low). The Blue Nile has experienced an overall decline of 5% in seasonal summer precipitation, although most of this decline has occurred since the mid-1970s. However, precipitation over the Blue Nile catchment in 1988 reversed this recent trend, with precipitation 13% above the century average. Although of lesser significance for Nile discharge, precipitation yields over central Sudan are an important indicator of the behaviour of the African monsoon. Here, there has been a decline of nearly 10% in seasonal summer precipitation over the twentieth century, with most of this reduction having occurred since the mid-1960s. Precipitation values for 1988 and 1989 have not recovered to the century average. These recent changes in central Sudan represent up to 25% of the precipitation fluctuations experienced during the Holocene period.

Likely precipitation levels over the Nile Basin in the future are difficult to establish. One method presented here uses a climate scenario developed from a number of GCM experiments. This scenario (which may be realized by the second half of the twenty-first century) suggests mean seasonal temperature increases over the Nile Basin of between 2° and 4°C. This increase might be slightly greater over the Lower Nile Basin than over the Upper Nile Basin. There is no clear signal from the composite model scenario about the magnitude or extent of precipitation changes over the Blue Nile catchment or over central Sudan. There is a slight tendency for reduced summer precipitation over the Blue Nile and increased winter precipitation over central Sudan. Over the White Nile catchment the scenario suggests a high probability (between 0.8 and 0.9) for increased precipitation yield in both summer (+18%) and winter (+28%) seasons.

With the inevitable increase in potential and actual evapotranspiration which would result from the higher surface air temperatures, combined with the scenario precipitation projections, our current best guess for greenhouse-induced forcing of Nile discharge would be for reduced Blue Nile flows and constant or slightly increased White Nile flows. Such a suggestion is not wholly incompatible with the observed precipitation trends over the White and Blue Nile catchments in recent decades. Obviously, this scenario remains highly uncertain until substantial improvements are made in the modelling of global climate using the new generation of combined ocean-atmosphere GCMs.

Two implications follow from the analysis presented in this chapter. First, the magnitude of historical fluctuations in Nile discharge (experienced both in this century and in previous millennia) is unlikely to be reduced in the future. Indeed, the additional climate-forcing which will result from increased greenhouse gas concentrations may well increase the amplitude of such fluctuations and may also generate a quite different seasonality in the contributions of the Blue and White Niles to main Nile discharge. Second, the greater confidence we have in the model projections of increased temperatures than in the projections of altered precipitation over the Nile Basin suggests that increased water loss through increased evapotranspiration should preoccupy our attention. Part, or even all, of this water loss may be compensated by increased precipitation yield over the White Nile catchment, but either possibility lends greater emphasis to the need for the completion of the Jonglei Canal and other planned water-conservation projects in southern Sudan in order to lessen the vulnerability of the Nile waters to evaporation loss.

## Acknowledgements

Part of the work on extending and updating the Global Precipitation Dataset was undertaken as part of contract PECD/7/10/198 funded by the Department of the Environment. The GCM climate scenario for the Nile Basin was supplied by Ben Santer of the Max Planck Institute, Hamburg, and the temperature curves in Figure 17.5 by Phil Jones.

## References

Adamson, D. A., Gasse, F., Street, F. A. and Williams, M. A. J. (1980) 'Late Quaternary history of the Nile', *Nature*, **288**, 50–55.
Ahlcrona, E. (1988) *The Impact of Climate and Man on Land Transformation in Central Sudan*, Lund University Press, Sweden.
Bolin, B., Doos, B. R., Jager, J. and Warrick, R. A. (eds) (1986) *The Greenhouse Effect, Climate Change and Ecosystems*, SCOPE Vol. 29, Wiley, Chichester.
Bradley, R. S., Diaz, H. F., Eischeid, J. K., Jones, P. D., Kelly, P. M. and Goodess, C. M. (1987) 'Precipitation fluctuations over northern hemisphere land areas since the mid-19th century', *Science*, **237**, 171–175.
CLIMAP (1976) 'The surface of the ice-age earth', *Science*, **191**, 1131–1137.

COHMAP (1988) 'Climatic changes of the last 18 000 years: observations and model simulations', *Science*, **241**, 1043–1052.

Currie, R. G. (1987) 'On bistable phasing of 18.6 year induced drought and flood in the Nile records since AD 650', *J. of Climatology*, **7**, 373–390.

Diaz, H. F., Bradley, R. S. and Eischeid, J. K. (1989) 'Precipitation fluctuations over global land areas since the late 1800s', *J. of Geophys. Res.*, **94** (D1), 1195–1210.

Farmer, G. (1989) 'Rainfall', in *The IUCN Sahel Studies, 1989*, IUCN, Gland, Switzerland.

Griffiths, J. F. (ed.) (1972) *Climates of Africa*, Vol. 10 of *World Survey of Climatology*, Elsevier, Amsterdam.

Hassan, F. A. (1981) 'Historical Nile floods and their implications for climatic change', *Science*, **212**, 1142–1145.

Hulme, M. (1989) *Analysis of worldwide precipitation records and comparison with model predictions*, Report for DoE Contract PECD/7/10/198, September.

Jones, P. D., Wigley, T. M. L., Folland, C. K. and Parker, D. E. (1988) 'Spatial patterns in recent worldwide temperature trends', *Climate Monitor*, **16**, 175–185.

Nicholson, S. E. (1988) 'Land surface–atmosphere interaction: physical processes and surface changes and their impact', *Prog. in Phys. Geog.*, **12**, 36–65.

Parker, D. E., Folland, D. K. and Ward, M. N. (1988) 'Sea-surface temperature anomaly patterns and prediction of seasonal rainfall in the Sahel region of Africa', in Gregory, S. (ed.), *Recent Climatic Change*, Belhaven Press, London.

Sutcliffe, J. V., Dugdale, G. and Milford, J. R. (1989) 'The Sudan floods of 1988', *Hydr. Sciences Journal*, **34**, 355–364.

Wickens, G. E. (1975) 'Changes in the climate and vegetation of the Sudan since 20 000 BP', *Boissiera*, **24**, 43–65.

Wickens, G. E. (1982) 'Paleobotanical speculations and Quaternary environments in the Sudan', in Williams, M. A. J. and Adamson, D. A. (eds), *A Land between Two Niles*, Balkema, Rotterdam.

Williams, M. A. J. and Faure, H. (eds) (1980) *The Sahara and The Nile*, Balkema, Rotterdam.

Wigley, T. M. L., Santer, B. D., Schlesinger, M. E. and Mitchell, J. F. B. (1990) 'Developing climate scenarios from equilibrium GCM results' (in press).

# 18 EFFECTS OF CLIMATIC CHANGE ON WATER RESOURCES FOR IRRIGATION

Wulf Klohn
*FAO, Rome, Italy*
and
Nigel Arnell
*Institute of Hydrology, Wallingford, UK*

## 18.1 Introduction

Many important decisions in development planning and in the design of engineering projects, including irrigation, agricultural land use and drought relief, are based on the assumption that past climate data are a reliable guide to the future. In recent years, the international scientific community has reached a consensus that the climate is changing towards global warming. The main cause for climate change is the increasing concentration, owing to human action, of carbon dioxide ($CO_2$) and other greenhouse gases in the atmosphere. Climate and hydrological data can no longer be seen as stationary, and it is necessary to refine estimates of future climatic conditions and of the resulting hydrological regime to improve the quality of decisions (WMO, 1986).

There is hardly any aspect of the biosphere, including agriculture, that will not feel the impact of climate change. There is a general consensus that there will be a substantial global surface warming, with amplification towards the poles. The magnitude and speed of changes cannot as yet be quantified with confidence. Regarding precipitation and soil moisture at any given location, not even the sign of change can be given with any degree of reliability (Bach, 1989; Beran, 1989). The odds are, however, that there will be larger variability of climate, i.e. that extreme events, such as floods and droughts, will become more frequent, and that there may be less resources available in many arid regions. Further converting climate information to water-resource decisions involves many additional uncertainties due to the complexity of practical cases. It is probable that another ten years or so may elapse before sufficient data will have accumulated, and climate model results will have converged sufficiently to validate adoption of new analytical procedures and possibly to justify alterations to existing facilities (WMO, 1989). It is nevertheless not too early for those involved with water resources to include the certainty of variability and the possibility of change in their thinking about future water provision.

This chapter reviews aspects of relevance to the Food and Agriculture Organization of the United Nations (FAO) and discusses the possible impact of climate change on agricultural water management.

## 18.2 Climate change and FAO

FAO has intergovernmental responsibilities in the field of agriculture and rural development. In this context, the term 'agriculture' encompasses the related natural resources, particularly soils and water, and also includes animal husbandry, fisheries, forestry and wildlife.

Climate change is a priority issue for FAO. Among the matters of special concern to the Organization are the links between climate change and deforestation and the possible role of reforestation in slowing down climate change or mitigating its impact; the implications of sea-level rise for food production in low-lying areas; potential shifts in agro-ecological zones that could alter positively or negatively, depending on the country, cropland availability and suitability; and the impact of changes in total rainfall and its greater variability on food production and agricultural export earnings.

In the fulfilment of its tasks, the Organization has developed over the years a number of monitoring and planning tools. Those that can be seen as most important to monitor climate change and its consequences are:

1. A world agriculture information centre which maintains and continuously updates information on agriculture, livestock, fisheries and forestry;
2. A Global Information and Early Warning System which allows in particular for timely response to food and agriculture emergencies;
3. An agrometeorological unit which collects and disseminates climatic information for planning and development in the field of agricultural production;
4. A remote-sensing programme for collection and interpretation of earth resources and meteorological satellite data in support of early warning services related to agricultural production and food security.

Within the framework of its mandate, the Organization carries out, or envisages to carry out, the following activities related to climate change:

1. Maintaining a watching brief on climate modelling, and monitoring and assessing actual or potential impacts on crops, livestock, fisheries and forestry production;

2. Development of policies and technologies to increase resilience of production systems to inter-annual and within-season climatic variability;
3. Intensification of early warning and disaster-prevention activities with respect to increased frequency of floods and droughts;
4. Investigation of potential changes in crop distribution and productivity in the Sahel;
5. Strengthening and harmonization of international cooperation to reduce deforestation and speed up agroforestation and reforestation so that forests in the long term serve globally as a sink for carbon dioxide;
6. Examination of tree planting and natural forest management policies and practices to mitigate emission of carbon dioxide, leading also to better forecasts of the climate situation;
7. Global reassessment of the situation and trends of tropical forest resources with 1990 as a reference year, applying the same concepts and classification used in the FAO/UNEP assessment carried out ten years ago.

As a result of these activities, the Organization expects to make available to its members and the international community relevant information on the global aspects of climate change, so as to facilitate appropriate political action. At the same time, FAO aims at making available to its members, and in particular to developing countries, proven technology and know-how to assess and mitigate the regional and local consequences of climate change.

## 18.3 A case study

One-third of the world's food is grown on irrigated land and a change in climate which alters the ability to irrigate would have very significant regional and global impacts on food production and security. It is difficult to generalize the impacts of change on irrigation, however, because these impacts are quite sensitive to the nature of the climatic change – changes in the date of onset of the wet season, for example – and assumptions made about the characteristics and efficiency of an irrigation scheme and the response of its operators to changed conditions.

Alterations in irrigation potential can only therefore be determined through case studies which combine information on future climatic and hydrological characteristics with a realistic irrigation demand and management model. The FAO commissioned such a study from the Institute of Hydrology (IH) in 1987, with the primary objective of developing a methodology for impact assessment and identifying the critical elements in the process.

A water resources study of the 3240 km$^2$ Malibamatso catchment in Lesotho in Southern Africa had recently been undertaken by IH (as part of a larger consortium), and a simple rainfall-runoff model had been calibrated to estimate monthly river flows from monthly rainfall and evaporation inputs. This model could easily be used to estimate the consequences of changes in climatic inputs, and accordingly a case study was devised using a hypothetical irrigation scheme in the Malibamatso catchment. A hypothetical – but realistic – crop pattern was defined, and crop water requirements were computed using FAO methodology (FAO, 1977). A simple soil water balance model was used to compute the amount of rainfall available for crops. For part of the analysis it was assumed that a reservoir, with a fixed storage capacity, supplied water to an irrigation scheme.

The hydrological and irrigation models were first run using 56 years of observed monthly rainfall data as input and assuming that evaporation was the same each year. These climatic inputs were perturbed in subsequent runs, according to several scenarios of future change. One group of runs involved arbitrary changes of precipitation of plus or minus 10% and temperature increases, throughout the year, of 1°, 2° or 4°C. Another simulation run used a climate change scenario derived from a general circulation model (GCM): temperatures were assumed to be 6°C higher throughout the year, and precipitation was assumed to increase during the dry season but decrease in the summer wet season. It must be emphasized that such a climate change is not a forecast of future conditions; it is merely a statement of one possible future (see Beran, 1989). In these pilot studies, no attempt was made to account for changes in plant productivity or water use in a high-$CO_2$ atmosphere.

Three broad aspects of irrigation scheme performance were considered in the study, namely, sensitivity of irrigation demand to changes in precipitation and evaporation, in the reliability of the irrigation scheme, and in the areas which can be irrigated with a given degree of reliability.

Mean annual irrigation demand was found to be most sensitive to changes in evaporation. A change in precipitation of 10% gave a change of approximately 5% in irrigation demand, while an increase in evaporation of 10% (corresponding to an increase of around 2°C) increased irrigation demand by 18%. These sensitivities depend, of course, on the seasonal distribution of climatic change, but give an indication of the very significant impacts of change on irrigation requirements: demands increased by over 60% under the GCM scenario.

The reliability of the hypothetical irrigation scheme was assessed by determining the proportion of years in which the amount stored in the reservoir was insufficient to meet demands. Scheme reliability was found to be very dependent on climatic inputs. Under current conditions, an irrigation scheme covering 35 700 ha would receive insufficient water in one year in five: if temperatures were to increase by 2°C (with no change in precipitation) such a scheme would fail in every other year, and under the GCM-based scenario would fall in virtually every year. The area of scheme that could be supplied with the current level of reliability would be much smaller under a drier, warmer climate.

A more sophisticated analysis allowed the cropped area to be reduced as a water shortage developed, and the mean cropped areas, for each crop, over several years were calculated under all the assumed climates. These calculations could provide the basis for estimates of future average and extreme food production.

The detailed conclusions and sensitivities inferred from the case study are, of course, dependent not just on the climate change scenarios used but also on the current climatic, hydrological and management characteristics of the scheme. However, several general methodological conclusions were drawn from the study. First, the response of an irrigation scheme to climate change can be very sensitive to the assumed nature of that change, and in particular to the seasonal dimensions of change. It is not possible, given the current state of GCM modelling (Bach, 1989), to make definitive statements of the future climate in a region, and any impact study must therefore consider several different – but feasible – scenarios. Second, the impact of change on irrigation demand and feasibility may be difficult to infer from changes in climate and hydrology alone. The timing of cropping, for example (which will be determined by crop types planted), will influence changes in demands. Finally, any evaluation of future changes must incorporate an assessment of possible management responses to water shortage and climate change: how would cropping patterns change under conditions of more frequent water shortage, for example, or how would scheme efficiencies alter? The case study confirms, however, that it is currently possible to derive some useful insights into the sensitivity of irrigation schemes to future climate change.

## 18.4 Climate change impact assessment through agro-ecological zoning

The agro-ecological zones (AEZ) and potential population capacity studies carried out by FAO had two main objectives; first, to estimate the physical potential of given land areas to produce food and, by reference to per capita food requirement, to support population; and, second, to compare the population which could be supported in given land areas with known and projected populations. In summary, the methodology consisted of overlaying a climatic inventory on a soil map so as to identify areas with quantified soil and climatic conditions important to crop production. Each area of similar soil and climatic conditions was termed an agro-ecological cell. The production potential of each cell was then estimated in terms of specific food crops under defined levels of inputs. Finally, potential production under these different assumptions and for present and projected time periods was converted into food nutrients and, by reference to per capita human food requirements, to the physical potential of land resources to support present and projected populations (Higgins *et al.*, 1986). The assessment helped to focus attention by the international community and of governments concerned on countries facing a critical food situation in the foreseeable future.

The inherent AEZ methodology lends itself to assessment of the consequences of climate change on a continental and global scale by introducing climate inventories adjusted to reflect climate change scenarios emerging from global atmospheric circulation models and other climate-projection techniques. Besides altering the climatic regions used for the AEZ study, it should also be possible to refine the climatic assumptions by introducing further probability concepts.

The projections of irrigation production used in the AEZ study, and other projections, do not consider the availability of water as a limiting factor to the potential population-supporting capacity. However, a change in climate, as suggested by the results of the Lesotho case study, is likely to require more water to sustain the same food production in irrigated agriculture. The outlook on food self-sufficiency could consequently be considerably modified. The analysis of water resources as impacted by climate change under the same scenarios provides a possibility to indicate whether a threat to food security does exist, and if so, to what extent it can be mitigated by irrigation and water management practices.

Work on an update of the AEZ study in the context of climate change and limited water availability is still on the drawing board. One may, however, try to work out the general direction of its results. One obvious case is Egypt, with almost zero rainfed agricultural potential. At the global scale, roughly two-thirds of irrigated land is in Asia, and AEZ has shown that in Asia there is limited surplus food-producing potential without irrigation. Irrigation in Asia is already largely concerned with the production of staple food crops. Taking into account the population projection for this part of the world, it is certain that increased water requirements for irrigation due to climate change could pose a food security threat (Nemec, 1989). Overall efficiency of water use in irrigation is still low and allows some leeway. One may therefore conclude that management practices of water for irrigation need to move into a higher level of efficiency and accuracy to maintain an adequate food supply to an increasing population.

## 18.5 Conclusions

The effects of climate change on water resources available in a particular region cannot currently be predicted with any degree of confidence. Besides higher surface temperature and evaporation (if humidity is available), it is generally assumed, in descriptive terms, that there will be higher variability (more floods and droughts) and that regions where a high proportion of available water is already utilized may be more severely impacted. While awaiting the climate-monitoring and research results which will give a better understanding of practical problems which may arise, governments, planners and operators of water resources systems require advanced warning on the potential problems, and this should be founded on the best available factual base. Advice on large-scale and long-term planning, as needed by the international community and by countries with large and complex territories, could be substantiated through application of the agro-ecological zoning methodology using climate scenario data and general indices of water resources availability. At the local scale, practical situations are so complex that assessments and remedial action are likely to be site-specific. The case study discussed in this chapter points to a strong possibility that, under changed climate conditions, more water will be required to keep the same level of production from irrigated agriculture. Consequently, any strategies developed to promote and guide international cooperation relating to the possible impact of climate change on irrigated agriculture should focus on availability of proven operational technology to support national managers and technicians in achieving better management of agricultural water resources. Water resources systems whose behaviour is well understood and which are closely monitored and rationally managed have a better chance of accommodating to changing conditions. It is therefore concluded that the same measures needed to improve current irrigation water control and use efficiency provide also the best guarantee to cope with the effects of climate change on irrigated agriculture.

## Acknowledgements

The Lesotho case study was carried out by Ben Piper, Max Beran and Nigel Arnell at the Institute of Hydrology, Wallingford, UK.

## References

Bach, W. (1989) 'Projected climate changes and impacts in Europe due to increased $CO_2$', in *Conference on Climate and Water*, Helsinki, 11–15 September, Government Printing Center, Helsinki.

Beran, M. (1989) 'The impact of climatic change on the aquatic environment', in *Conference on Climate and Water*, Helsinki, 11–15 September, Government Printing Center, Helsinki.

FAO (1977) 'Crop water requirements', Irrigation and Drainage Paper No. 24, Rome.

Higgins, G. H. *et al.* (1986) 'The FAO agro-ecological zones approach to quantification of land resources and assessment of crop and population potentials', Workshop on Agro-ecological Characterization, Classification and Mapping, CGLAR-FAO, Rome.

Higgins, G. M., Dieleman, P. J. and Abernethy, C. L. (1988) 'Trends in irrigation development and their implications for hydrologists and water resources engineers', *J. Hydr. Sciences*, **33**, 1, 2.

Institute of Hydrology (1988) *Effects of climatic change on water resources: an example from Lesotho*, report prepared for FAO, Wallingford.

Nemec, J. (1989) 'Impact of climate variability and change on water resources management in agriculture', in *Conference on Climate and Water*, Helsinki, 11–15 September, Government Printing Center, Helsinki.

WMO (1986) *Report of the International Conference on the Assessment of the Role of Carbon Dioxide and of other Greenhouse Gases in Climate Variations and Associate Impacts*, WMO No. 661, Geneva.

WMO (1989) *Statement on the Hydrological and Water Resources Impacts of Global Climate Change*, EC–XXXXI, Geneva.

# 19 SOCIAL CONSTRAINTS ON ADJUSTING WATER RESOURCES MANAGEMENT TO ANTHROPOGENIC CLIMATE CHANGE

William E. Riebsame
*University of Colorado, USA*

## 19.1 Introduction

If the climate should change over the next several decades due to human enhancement of the global greenhouse effect, what options would water resource managers and planners have to adopt? This question is compelling because of the increasing credibility of global warming projections (see, for example, World Meteorological Organization, 1985, 1988). It is also of fundamental interest to students of water development because it touches on basic questions: How robust and resilient are development plans and current systems? How does their capacity to absorb shocks change through time? How is this related to the conomic, social, and cultural environment in which resources management is embedded? What will be the nature of future water development – with or without global warming?

Water resources development reflects overall social development, and manifests peoples' desires for a better life. It mirrors economic, geographical and political trends, and binds the fortunes of diverse localities, populations, and countries (Baumann and Haimes, 1988). Sewell and White, discussing the prospects for integrated development of the Mekong, wrote in 1966 that:

International river basin development will undoubtedly be one of the major means of accomplishing economic growth and social change in the developing countries. Most of the world's major rivers are international rivers, and most flow through the developing countries. Approximately 150 river basins straddle international boundaries, and together they cover almost half of the world's land surface, excluding Australia and Antarctica. Some of these rivers could be utilized for the production of hydroelectric power, the provision of water for irrigation or domestic and industrial uses, the improvement of navigation, or the control of floods (1966, p. 5).

This statement still represents a manifesto for contemporary river basin development. Yet, development efforts around the world have met with mixed success, especially in the least-developed regions. Projects occasionally cause new social and environmental problems: population dislocation, disruptive land-use change, and inequities in the distribution of resources. Yet significant regional water developments are still being planned in many areas, and atmospheric researchers are adding a credible threat of climate change to the list of problems that might disrupt those developments.

## 19.2 Global warming: ultimate threat or typical problem?

In the complex technical and social realm of water development, does global warming pose a unique threat or is it simply another factor to be accommodated, more or less successfully, along with changes in population, demand, funding, policy, etc.? Perhaps because we have addressed the question of climate change impacts seriously now for only about 15 years, there is no widely accepted answer to this question. Instead, we have assessments suggesting serious impacts in many cases, large uncertainty, and something of a debate between optimistic and pessimistic analysts on the threat that global warming poses.

Discussions of global warming impacts tend to bifurcate between two fundamental precepts about the resiliency and adaptability of resource management systems. One group of analysts points out that the rate of climate change predicted by global climate models (GCMs) is larger than any experienced in human history (e.g. Schneider, 1989). This is taken, *ipso facto*, and with reference to recent climate impacts (e.g. extended Sahelian drought) to mean that global warming will be disruptive and painful, and poses the chief environmental threat of our times. Essentially, all the current concern over the issue stems from the intuitive sense that resource systems are tuned to the current climate and will suffer under any significant change (though analysts do occasionally suggest that global warming will yield both 'winners' and 'losers').

The other precept holds that society is quite resilient, and that, with some exceptions, food, water, energy and other resource systems are robust and adaptable, and can be adjusted (or will 'automatically' adjust) apace climate change (e.g. Wittwer, 1980; UN Environment Programme, 1987). The exceptions, of course, are the poorest countries – societies lagging in development – or a few regions where the physical effects are especially catastrophic (e.g. low-lying lands subject to sea-level intrusion).

Unfortunately, there is only a limited research base for an evaluation of the overall vulnerability or resiliency of modern resource systems or their near-future incarnations. It is perhaps in water resources planning that vulnerability and resilience concepts are best developed, though they are still difficult to operationalize beyond a simplified system (e.g. a single reservoir), and do not apply to such an inclusive concept as 'river basin development'. Yet it is precisely the integrated aspects of water and associated resource development that must be understood better to assess the threat of climate change.

## 19.3 Water system impacts and resilience

Emerging research on water resource impacts of climate change tends to indicate that even small climate changes could yield large changes in raw water supply (e.g. Revelle and Waggoner, 1983; Gleick, 1986). Most analysts argue that modern water management systems are capable of absorbing significant climate change without failing, chiefly due to the risk-averse nature of water system engineering and operation (see, for example, Hanchey *et al.*, 1988; Georgeson, 1986). Water project designs generally incorporate liberal safety margins to accommodate errors and unusual conditions (Schwartz, 1977; Cohen, 1986; Schilling *et al.*, 1987). Indeed, Hanchey *et al.* (1988) argued that water project planners had

> over the course of fifty years of application and refinement [developed] a large body of empirical and theoretical procedures and decision rules that have yielded what are generally considered to be fairly robust and resilient project designs . . . This empirical approach, emphasizing as it does extremes of climate variability over the past 100 years, encompasses a significant proportion of the anticipated [climate] changes, at least for large scale water management (p. 399).

This risk-averse design approach considers relationships between critical climate variables (e.g. runoff) and water project operating limits (Figure 19.1(a)). Designers plan for extreme events, within technical and economic limits, and add additional safety margins according to rules of thumb and the overriding principle that the system simply *must not fail* (Georgeson, 1986; Dziegielewski, 1986). The net social costs and benefits of this strategy are unknown (Schilling *et al.*, 1987), though it has indeed produced mostly reliable water projects when sufficient resources and will were available to carry out project design.

Yet in terms of broad development goals, water projects in some regions have failed to provide reliable outputs, or benefits have barely outpaced the losses (be they economic, social, or environmental), especially in large-scale developments. Even where projects are functioning well, the simple fact remains that climate change is, by definition, a change in the

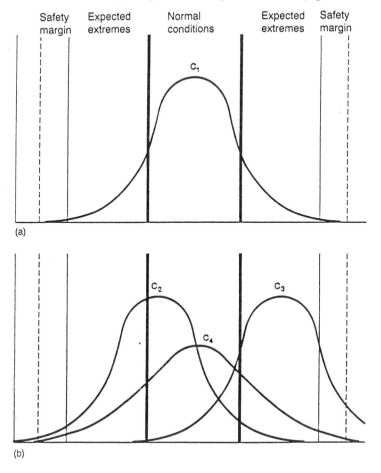

**Figure 19.1**

statistical distribution of climatic elements. Depending on the safety margins designed into a project, climate change will alter the frequency of conditions that approach or surpass failure thresholds (Figure 19.1(b)). If planners have achieved socially acceptable project reliability, then any climate change larger than the uncertainty inherent to hydroclimatological analysis violates explicit and implicit planning criteria.

Experience suggests that water systems exhibit a range of sensitivities to shocks such as climate fluctuations, both among systems and in a given system over time. They also encompass a range of potential purposeful and incidental adaptations to climate through changes in physical facilities, operations, or demand, narrowed or enlarged by social factors (Riebsame, 1988a,b). While it is a straightforward process to calculate the size of reservoirs needed to maintain yield reliability under a new climate, or to design any operational rules that accommodate altered climate inputs, the ability of planners and managers to link climate change to planned adjustment is more difficult to assess.

## 19.4 Constraints on water resource adaptation

The remainder of this chapter describes a set of constraints or problems that theory and experience suggest will come into play as resource managers grapple with climate change (Table 19.1). Examples are drawn especially from two river basins subjected to coordinated, integrated planning and development: the Sacramento Basin in the western USA, and the Lower Mekong in South-east Asia.

**Table 19.1**   Problems in adjusting water resources management to climate change

---

Detection/Interpretation
Operational Adjustability
Institutional Adaptability
Adaptive resource Availability
Policy Interference
Surprise: unexpectable and expectable

---

*19.4.1 Detection/interpretation*

Water resource planners are understandably hesitant to change their designs in anticipation of a future, uncertain change in climate. Moreover, there is good reason to expect that even rapid climate change will be difficult to detect early, and that ambiguous indicators will be interpreted in ways that delay adjustment, due to at least three problems:

1. *Climate noise*: the climate data will obscure global warming for several more years even at 'rapid' rates of global warming, and credible scientists will argue for and against a true 'signal' of global warming for some time to come. The latest assessments of global modelling indicate that regionally credible projections of global warming will not be available for the next 5–10 years.
2. *Impacts projection/detection*: the physical complexity of water systems and basin developments will obscure the early impacts of climate fluctuations. By definition, this might be taken to indicate that the problem is not yet sufficiently serious to warrant action (small impacts), but it also delays recognition of a trend that, according to greenhouse theory, will continue and probably accelerate once established.
3. *Individual and institutional cognitive dissonance*: behavioural research suggests that water planners, and other natural resource managers, expect *a priori* the large-scale environment to be stable (Holling, 1986; Morrisette, 1988), and are thus biased against recognizing and acting on indications of change. If true, this probably stems from a constellation of

factors including limits on individual decision making under uncertainty, institutional constraints, etc.

The last constraint is especially critical but less well studied. A rich body of research demonstrates that decision makers respond to environmental threats as they perceive them, not necessarily as they can be described, measured, or analysed by 'objective' criteria (Whyte, 1985). A 'stability bias' is especially strong in water resource planning where long-term future conditions are explicitly assumed to emulate those observed over the past several decades. Statistics describing climatic variables are assumed to be stationary over time, showing no cumulative trends, an expectation that pervades essentially all hydrological calculations (Changnon, 1984; Lettenmaier and Burgess, 1978; Riebsame, 1988a–c).

Of course, absent detailed and reliable forecasts of future climate or empirical evidence of trends, assumed climatic stationarity makes sense. However, increasingly credible predictions of global climate provide a rationale for at least examining management practices in the light of the *potential* for climate change, and for exploring the role of climate perception in natural resource planning.

Indeed, managers may be forced by the long planning horizons associated with resource systems, and by growing public pressure, to assume that the climate is changing before predictions become much more detailed, and before there exists incontrovertible evidence for actual climate change. In this ambiguous situation, beliefs and attitudes about climate and climate change are likely to play an important role in planning decisions.

Climate perception surveys are rare, but in a recent survey of water resource planners in the south-western and south-eastern United States it was found that while all were at least aware of the global warming issue, few expected to experience actual climate change in the foreseeable future, and few felt that they would be forced to change their management practices due to climate change in the next three decades. As the threat emerges, those who wish to evoke action face a large educational and information transfer task.

## 19.4.2 Operational adjustability

Water systems exhibit differing capacity for operational adjustment as climate changes. This depends on both physical and policy flexibility, and we lack methods for measuring and comparing this quality of water systems. Operational adjustments occur at the annual to decadal time scale, and are affected by:

1. *Range of options*: the range of choices available to or recognized by water systems managers; their perceived 'response repertoire';

2. *Automatic adjustments*: the potential for automatic adjustment inherent in the existing 'operational canon' (e.g. 'rule curves' based on climate conditions or as operational criteria such as project floods are recalculated over time);
3. *Emergency response*: ability to take extraordinary measures under new and unusual conditions treated as special or short-term emergency situations.

### 19.4.3. Institutional adaptability

At the time scale of years and decades a key element of systemic resilience is the adaptive capacity of the social institutions that guide water management. Among these institutions are legal structures (water laws, treaties, compacts), agencies responsible for development and routine management, and the broader social structures in which water management is practised – from professional to political.

The emerging literature on institutional change offers cause for both optimism and pessimism on this score. It ranges from claims that water institutions themselves have shaped the larger social environment in a despotic fashion (Wittfogel, 1957) to case studies showing that agencies do change, if slowly, in response to broader social change – even institutions cited as exemplars of rigidity.

Legal institutions may pose more enduring obstacles to adaptation. It is widely held in western US water analysis that the 'prior appropriation' and 'beneficial use' doctrines encourage inefficiency (e.g. Getches, 1988). Rigid inter-regional and international compacts or treaties emerge inevitably where contention and conflict are greatest, where systems are more sensitive to disruption. Analysts frequently cite the Colorado River compact, and its fixed allocation between the upper and lower basin, as limiting flexibility for dealing with problems such as salinity and climate change (Brown, 1988). Even relatively small changes in climate could cause serious water shortages in the Colorado Basin, where supply and demand are closely balanced (Revelle and Waggoner, 1983). A similar situation exists in the Indus, where there is growing conflict over upper and lower basin (as well as international) water allocation.

More research is needed on water institutions, on whether they exhibit similar degrees of rigidity or flexibility as managerial institutions. During the 1976–1977 drought in the Sacramento Basin, individual consumers showed great flexibility in terms of conservation, but the formal water users association, which represents the SWPs largest contract users, demanded maintenance of guaranteed contract deliveries.

### 19.4.4 Adaptive resources

It is often assumed that ability to adjust a water system or plan to accommodate changed conditions is chiefly constrained by financial limits.

This is certainly true in some cases, but may not be as pervasive as generally believed. Both the Mekong Committee and California agencies are experiencing declining water development budgets (Figure 19.2), but the Mekong, at least, can reasonably expect to obtain the international funding it requires to build the cascade of eight main-stem dams now planned, especially with growing Japanese involvement in development aid. California's SWP, and many urban water systems in the USA (e.g. Boston, Denver, Los Angeles, Tucson), are financially prepared to build new facilities, even with reduced federal assistance (see, for example, Schilling *et al.*, 1987). Most developments have been delayed by environmental and social concerns, not by money.

Funding constraints might come in competition between subsectors of resource systems (irrigation versus municipal), or as impacts draw resources to specific areas.

### 19.4.5 *Reduced tolerance of environmental and social impacts*

It can be argued that the recent history of water resources planning is dominated not by technical or financial hurdles and solutions but by growing intolerance of environmental and social impacts (changes). Development projects in both the Sacramento and the Lower Mekong basins are

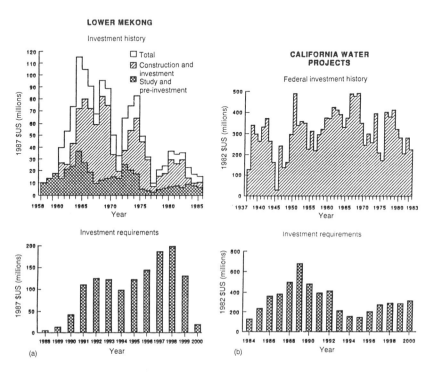

**Figure 19.2**

either on hold or have been scaled down or substantially altered in the last decade due to heightened concerns for environmental impacts. In the Scaramento this is illustrated by the Auburn Dam and Delta peripheral canal, facilities that would have increased the basin's firm yield by as much as 50% (California Department of Water Resources, 1987). In the Lower Mekong this constraint is reflected by changes in the scale of the planned cascade of dams, and the Mekong Secretariat's recent addition of environmental impact analysis as an integral planning element. Plans for the Pa Mong Dam, for example, have been considerably scaled down to reduce resettlement (from 250 000 to 50 000 people), despite marked reductions in the dam's ability to augment low flows – the basin's key problem and one of dam's chief relationales (Interim Committee for the Lower Mekong, 1988).

### 19.4.6 Policy interference

In any complex system there are negative feedbacks between elements and, occasionally, interfering goals. A classic example in water system planning that will affect climate change response is the antithetical nature of drought and flood protection. This policy interaction was seen in adjustment to recent climate extremes in the Sacramento Basin (Riebsame, 1988a–c). Maintenance of planned firm yield in the variable climate and rising demand (Figure 19.3(a)) that marked the 1970s and 1980s would have required increased storage capacity or reduced flood protection. With decreased political acceptability of large facilities, water managers sought a relaxation of flood regulations to free more reservoir space for conservation storage. They did not get it – instead, for one major reservoir the agency with flood-control jurisdiction (the Corps of Engineers) actually enlarged flood management capacity, reducing storage (Figure 19.3(b)). The SWP was forced to consider changing allocation rule curves that reduced the reliability of given water deliveries (Figure 19.3(c)).

Other potential policy interferences that might worsen in a changing climate occur at the intersections of treaty obligations and supply quantity and quality, or species protection and facilities development, especially as species are themselves affected by climate change. However, interference between two policy realms is difficult to anticipate, given the fragmented nature of water research and jurisdiction.

### 19.4.7 Surprise: Unexpectable and expectable

Finally, the global warming threat is especially fraught with the potential for surprise. Adaptation might be made easier or more difficult by conditions or situations that simply cannot be predicted. Global warming itself may yield unpredictable climate patterns or extremes to which no system can readily adapt. Social change and response might also surprise

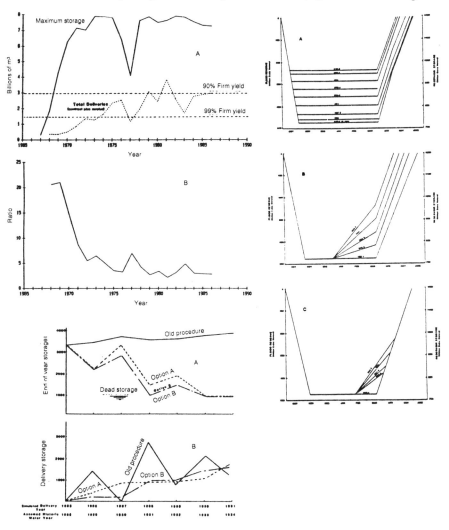

**Figure 19.3** *(a) Maximum annual storage, total deliveries and firm yield estimates with 1980 facilities (A) and ratio of storage to delivery (B) for the California State Water Project (source: California Department of Water Resources). (b) Flood-control diagram for the Folsom dam and reservoir showing adjustments through time: (A) original design, (B) 1977 change and (C) 1987 change. The diagram specifies allowable reservoir levels at different times of the flood season based on previous basin precipitation. The times with labels running from 215.9 to 533.4 (mm) represent basin wetness based on precipitation recorded during the previous six weeks (source: US Army Corps of Engineers, Mid-Pacific Region, Sacramento, CA). (c) Simulated SWP operations based on the 1977 rule curve and two alternatives proposed in 1985, for a hypothetical drought beginning with 1985 precipitation and storage conditions and following the pattern of the 1929–1934 design drought. (A) Total projected storage at the end of each simulated year; (B) delivery shortfalls from contract amounts (source: California Department of Water Resources)*

water planners: efforts to reduce greenhouse gases might reinforce support hydropower development or reduce demand for rice paddy inundation (two actions that would reduce emissions of greenhouse gases).

It may be useful, however, to differentiate between 'unexpectable' and 'expectable' surprise. This seeming contradiction recognizes that while we may face true surprises in a warming world (e.g. conditions that cannot be anticipated given current social and physical science knowledge), experience suggests that many of the surprises we are likely to encounter will stem more from failure to monitor or attempt to anticipate changing conditions than from lack of predictive ability. Post-audits of some recent resource management surprises, such as the reservoir system overload experienced on the Colorado River in 1983 (Rhodes *et al.*, 1984), or the large 1988 fires in Yellowstone National Park, indicate that the surprising conditions could have been anticipated with current knowledge.

## 19.5 Global warming as a water resources planning conundrum

The threat of global warming undermines the conventional water resource planning assumption that, while social factors (e.g. population) may change significantly, basic environmental elements such as climate are essentially stable. Because global warming is anthropogenic – stemming mainly from energy and agricultural activities – it also raises the question of how resource development plans contribute to (or mitigate) the problem.

### 19.5.1 Pressure to respond

High-level policy makers are responding to arguments for quick action to stem greenhouse gas emissions and to prepare for climate change, even without further improvements in scientific understanding of the problem – as evident in moves towards an international treaty to protect the climate (UN Environment Programme 1987, 1988; World Meteorological Organization, 1988). The public, elected officials, and researchers increasingly appreciate the potential impacts of global warming and the relationship between policy delays and worsened future impacts (Jager, 1988). Most resource managers, however, have adopted a 'wait-and-see' posture on global warming, and are attracting criticism for failing to address the issue aggressively (US Senate Agriculture Committee, 1989; *New York Times*, 1989).

A wait-and-see approach is justified chiefly by three arguments: climate change predictions are too uncertain, especially at the regional level, for specific action (Katz, 1988); current systems can absorb significant climate change without failing (Hanchey *et al.*, 1988); and technological change can override the negative effects of climate change (Wittwer, 1980). All these arguments have some merit. Global warming predictions do lack the regional details necessary to alter development plans to accommodate climate change: it is premature to build new reservoirs or to plant different tree species because of the greenhouse threat. Yet current planning for basin development tends to neglect the logic of initial sensitivity studies based on climate scenarios. (Are there examples of where climate change is being considered?) Moreover, current planning policies and institutional arrangements dissuade managers from accounting for potential impacts on global commons such as the atmosphere and climate (Schelling, 1983).

Water planners face a conundrum: how to respond to the pressure for action in the face of large uncertainty without derailing desirable development or introducing unnecessary inefficiencies into systems.

## 19.5.2 An incremental response

A planning process that better accommodates the nature of the climate change problem is needed, one that reflects the sensitivity of water systems to climate change; uncertainty surrounding climate changes and how it can be incorporated into their response repertoire; and the role that resource activities play in worsening or lessening the greenhouse and other environmental problems. Future planning should incorporate at least four elements:

1. Better links between climate researchers and information and water planning;
2. Resource sensitivity analysis that explicitly recognizes the potential for fundamental environmental change;
3. Incremental adjustment linked to the increasing certainty of greenhouse effects or demonstrable impacts;
4. Consideration of an enlarged range of alternative adjustments in case significant climate change occurs (Table 19.2).

**Table 19.2** Approaches to adjusting water resources management to climate change

| |
| --- |
| Step-wise adjustment |
|   Sensitivity analysis |
|   Incremental adjustment as certainty increases |
|   Enlarge the range of alternatives |
| Reduce constraints on adaptation |
| Information provision (clearing house) |
| Create mechanisms for expanding the range of options |
| Expanded tie-in strategies |

First, managers should inform themselves about the issue, seeking advice and information from credible sources such as the World Climate Programme (and not the news media). They will find that the information contains various amounts of certainty (the world will warm) and uncertainty (the distribution and rate of change). Sufficient information, and credible climate scenarios, are now available for sensitivity analyses that tell a planner how soon the system might incur impacts from climate change.

Next, to avoid being pressured into action prematurely, or paralysed by uncertainty, managers should consider a wider repertoire of planning methods and adjustments than has been relied upon traditionally. They should initially identify adjustments that are easily and cheaply implemented and reversed as needed (e.g. more frequent evaluation of reservoir operating), and ones that maintain or expand, rather than limit, future options (e.g. floodplain land use that places less fixed investment at risk), while substantially meeting current goals. For example, Sacramento Basin water planners should differentiate between adjustments that can be effected under current system arrangements (e.g. modifying reservoir operations and changing local water supply plans), and those that require more lengthy implementation (e.g. renegotiating user contracts or enlarg-

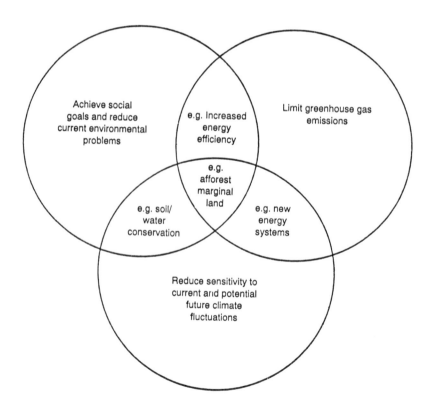

**Figure 19.4**

ing reservoir capacity). The social disruption, cost and irreversibility of adjustments must also be considered, and weighed against the risk of impacts as they become clarified.

Finally, resource planners should elaborate on the 'tie-in' strategy proposed by several climate and policy analysts (cf. Schneider, 1989), in which immediate steps to reduce greenhouse gas emissions are justified because they also solve current environmental problems such as acid precipitation. Additional 'tie-ins' are needed, such as between adjustments that not only reduce the global warming threat but make resource management systems less sensitive to *current* climate fluctuations *and* more adaptable to future climate change. These strategies coalesce in three realms, where (1) social needs and desires are met and current environmental problems are solved; (2) greenhouse gas emissions are limited; and (3) sensitivity to current and potential future climate fluctuations is reduced (Figure 19.4).

# References

Brown, B. (1988) 'Climate variability and the Colorado River Compact: Implications for responding to climate change', in Glantz, M. H. (ed.), *Societal Response to Regional Climatic Change: Forecasting by Analogy*, pp. 279–305, Westview Press, Boulder, CO.

California Department of Water Resources (1987) *California Water: Looking to the Future*, Bulletin 1860–87, State Printing Office, Sacramento.

Carbon Dioxide Assessment Committee (1983) *Changing Climate*, National Academy Press, Washington, DC.

Changnon, S. A., Jr (1984) 'Misconceptions about climate in water management', in Water Resources Center (ed.), *Proceedings of a Conference on Management Techniques for Water and Related Resources*, pp. 1–8, University of Illinois, Campaign-Urbana, IL.

Cohen, S. J. (1986) 'Climatic change, population growth, and their effects on Great Lakes water supplies', *The Professional Geographer*, **38**, 317–323.

Dziegielewski, B. (1986) 'Drought management options', in Water Science and Technology Board (ed.), *Drought Management and Its Impacts on Public Water Systems*, pp. 65–77, National Academy Press, Washington, DC.

Fischoff, B. (1981) 'Hot air: The psychology of $CO_2$-induced climatic change', in Harvey, J. H. (ed.), *Cognition, Social Behavior and the Environment*, pp. 163–184, Lawrence Erlbaum Associates, Hillsdale, NJ.

Georgeson, D. L. (1986) 'What are acceptable risks for public systems?' in Water Science and Technology Board (ed.), *Drought Management and its Impacts on Public Water Systems*, National Academy Press, Washington, DC.

Getches, D. H. (1988) 'Pressures for change in western water policy', in Getches, D. H. (ed.), *Water and the American West*, pp. 143–164, University of Colorado Natural Resources Law Center, Boulder, CO.

Glantz, M. H. (1982) 'Consequences and responsibilities in drought fore-casting: The case of Yakima, 1977', *Water Resources Research*, **18**, 3–13.

Gleick, P. H. (1987) 'Regional hydrological consequences of increases of atmospheric $CO_2$ and other trace gases', *Climatic Change*, **110**, 137–161.

Hanchey, J. R., Schilling, K. E. and Stakhiv, E. Z. (1988) 'Water resources planning under climate uncertainty', in The Climate Institute (ed.), *Preparing for Climate Change: Proceedings of the First North American Conference on Preparing for Climate Change: A Cooperative Approach*, pp. 394–405, Government Institutes, Inc., Washington, DC.

Hansen, J., Fung, I., Lacis, A., Rind, D., Lebedeff, S., Ruedy, R. and Russell, R. (1988) 'Global climate changes as forecast by Goddard Institute for Space Studies' three-dimensional model', *Journal of Geophysical Research*, **93**, 9341–9364.

Hansen, J. and Lebedeff, S. (1988) 'Global surface air temperatures: update through 1987', *Geophysical Research Letters*, **15**, 323–326.

Holling, C. S. (1986) 'The resilience of terrestrial ecosystems: local surprise and global change', in Clark, W. C. and Mann, R. E. (eds), *Sustainable Development of the Biosphere*, pp. 292–317, Cambridge University Press, Cambridge.

Hoyt, W. G. and Langbein, W. B. (1955) *Floods*, Princeton University Press, Princeton, NJ.

Interim Committee for the Coordination of Investigations of the Lower Mekong Basin (1988) *Perspectives for Mekong Development*, Mekong Secretariat, Bangkok.

Jager, J. (1988) 'Anticipating climate change', *Environment*, **30**(7): 12–15, 30–33.

Jones, P. D., Wigley, T. M. L. and Raper, S. C. B. (1987) 'The rapidity of $CO_2$-induced climatic change: Observations, model results and paleoclimatic implications', in Berger, W. H. and Labeyrie, L. D. (eds), *Abrupt Climatic Change*, pp. 47–55, Reidel, Dordrecht.

Kates, R. W. (1980) 'Climate and society: Lessons from recent events', *Weather*, **35**(1), 17–25.

Kates, R. W., Ausubel, J. and Berberian, M. (eds) (1985) *Climate Impact Assessment: Studies of the Interaction of Climate and Society*, John Wiley, New York.

Katz, R. W. (1988) 'Statistics of climate change: Implications for scenario development', in Glantz, M. H. (ed.), *Societal Response to Regional Climate Change*, pp. 95–112, Westview Press, Boulder, CO.

Lamb, P. J. (1987) 'On the development of regional climatic scenarios for policy-oriented climate impact assessment', *Bulletin of the American Meteorological Society*, **68**, 1116–1123.

Lettenmaier, D. P. and Burgess, S. J. (1978) 'Climatic change: Detection and its impact on hydrologic design', *Water Resources Research*, **14**, 679–687.

Mearns, L. O. (1988) 'Variability', in Smith, J. and Tirpak, D. (eds), *The Potential Effects of Global Climate Change on the United States, Vol. 2:*

National Studies, Chapter 17, US Environmental Protection Agency, Washington, DC.

Mearns, L. O., Katz, R. W. and Schneider, S. H. (1984) 'Extreme high-temperature events: Change in their probabilities with changes in mean temperature', *Journal of Climate and Applied Meteorology*, **23**, 1601–1613.

Morrisette, P. M. (1988) 'The stability bias and adjustment to climatic variability: The case of the rising level of the Great Salt Lake', *Applied Geography*, **8**, 171–189.

Murphy, A. H. and Katz, R. W. (eds) (1985) *Probability, Statistics, and Decision Making in the Atmospheric Sciences*, Westview Press, Boulder, CO.

National Council on Public Works Improvement (1987) *The Nation's Public Works: Report on Water Supply*, Washington, DC.

*New York Times* (1989) 'Senator finds hole in federal report', 2 January, 12.

Phillips, D. H. and Jordan, D. (1986) 'The declining role of historical data in reservoir management and operations', in American Meteorological Society (ed.), *Preprints of the Conference on Climate and Water Management*, pp. 83–88, Boston.

Revelle, R. R. and Waggoner, P. E. (1983) 'Effects of a carbon dioxide-induced climatic change on water supplies in the western United States', in Carbon Dioxide Assessment Committee (ed.), *Changing Climate*, pp. 419–431, National Academy Press, Washington, DC.

Rhodes, S. L., Ely, D. and Dracup, J. A. (1984) 'Climate and the Colorado River: The limits of management', *Bulletin of the American Meteorological Society*, **65**, 682–691.

Riebsame, W. E. (1988a) 'The potential for adjusting natural resources management to climatic change: A national assessment', in Smith, J. and Tirpak, D. (eds), *The Potential Effects of Global Climate Change on the United States: Appendices*, Appendix J, US Environmental Protection Agency, Washington, DC.

Riebsame, W. E. (1988b) 'Adjusting water resources management to climate change', *Climatic Change*, **12**, 69–97.

Riebsame, W. E. (1988c) *Assessing the Social Impacts of Climate Change: A Guide to Climate Impact Studies*, United Nations Environment Program, Nairobi.

Riebsame, W. E., Diaz, H. F., Moses, T. and Price, M. (1986) 'The social burden of weather and climate hazards', *Bulletin of the American Meteorological Society*, **67**, 1378–1388.

Rosensweig, C. (1985) 'Potential $CO_2$-induced climate effects on North American wheat-producing regions', *Climatic Change*, **7**, 367–389.

Russell, C. S., Arey, D. G. and Kates, R. W. (1970) *Drought and Water Supply*, Johns Hopkins University Press, Baltimore, MD.

Schelling, T. (1983) 'Climatic change: Implications for welfare and policy'. in Carbon Dioxide Assessment Committee (ed.), *Changing Climate*, pp. 449–482, National Academy Press, Washington, DC.

Schilling, K., Copeland, C., Dixon, J., Smythe, J., Vincent, M. and Peterson, J. (1987) *Report on Water Resources*, Public Works Series,

National Council on Public Works Improvement, Washington, DC.

Schlesinger, M. E. and Mitchell, J. F. B. (1985) 'Model projection of the equilibrium climatic response to increased carbon dioxide', in MacCracken, M. C. and Luther, F. M. (eds), *Projecting the Climate Effects of Increasing Carbon Dioxide*, DOE/ER-0237, US Department of Energy, Washington, DC.

Schneider, S. H. (1989) 'The greenhouse effect: Science and policy', *Science*, **243**, 771–781.

Schwartz, H. E. (1977) 'Climatic change and water supply: How sensitive is the Northeast?' in Panel on Water and Climate (ed.), *Climate, Climatic Change, and Water Supply*, pp. 111–120, National Academy of Sciences, Washington, DC.

Sedjo, R. A. (1989) 'Forests: A tool to moderate global warming?' *Environment*, **31**(1), 14–20.

Shugart, H. H., Antonosvsky, M. Ya., Jarvis, P. G. and Sandford, A. P. (1986) '$CO_2$, climatic change and forest ecosystems', in Bolin, B., Doos, B. R., Jager, J. and Warrick, R. A. (eds), *The Greenhouse Effect, Climatic Change, and Ecosystems*, pp. 475–521, John Wiley, New York.

United Nations Environment Programme (1987) *The Greenhouse Gases*, UNEP/GEMS Environment Library No. 1, Nairobi, Kenya.

US Senate Agriculture Committee (1989) *Hearings on Global Warming and its Impacts on Agriculture*, GPO, Washington, DC.

West, M. W. (1988) 'Dams and earthquakes: A shaky relationship', *Civil Engineering*, **58**(4), 64–67.

White, G. F. (1988) 'Global warming: Uncertainty and action', *Environment*, **30**(6), i.

Whyte, A. V. T. (1985) 'Perception', in Kates, R. W., Ausubel, J. E. and Berberian, M. (eds), *Climate Impact Assessment: Studies in the Interaction of Climate and Society*, pp. 403–436, John Wiley, New York.

Wigley, T. M. L., Jones, P. D., and Kelly, P. M. (1986) 'Warm world scenarios and the detection of climatic change induced by radiatively active gases', in Bolin, B., Doos, B. R., Jager, J. and Warrick, R. A. (eds), *The Greenhouse Effect, Climatic Change, and Ecosystems*, pp. 271–322, John Wiley, New York.

Wittfogel, C. A. (1957) *Oriental Despotism: A Comparative Study of Total Power*, Yale University Press, New Haven, CT.

Wittwer, S. (1980) 'Carbon dioxide and climate change: An agricultural perspective', *Journal of Soil and Water Conservation*, **35**, 116–121.

World Meteorological Organization (1979) *Papers Presented at the World Climate Conference*, Geneva.

World Meteorological Organization (1985) *Report of the International Conference on the Assessment of the Role of Carbon Dioxide and of Other Greenhouse Gases in Climate Variations and Associated Impacts*, Publication No. WMO 661, Geneva.

World Meteorological Organization (1988) *Proceedings of the Conference on the Changing Atmosphere: Implications for Global Security*, WMO Publication No. WMO/OMM 710, Geneva.

# 20 SOME MAJOR IMPLICATIONS OF CLIMATIC FLUCTUATIONS ON WATER MANAGEMENT

Mahmoud A. Abu-Zeid
*Water Research Center, Cairo, Egypt*
and
Asit K. Biswas
*International Water Resources Association*
*Oxford, UK*

## 20.1 Introduction

The issue of climatic change and its possible impacts on mankind has become a major topic in recent discussion. The possible global warming due to the greenhouse effect is now firmly on the world's scientific and political agenda. Numerous meetings are now being held all over the world on climatic change which are scientific, political or even pseudo-scientific in nature. Unquestionably, ozone depletion and global warming have primarily been responsible for raising the interest of the general public in environment issues to a level that has not been seen since the early 1970s.

Much has been said and written recently on the impacts of potential climatic changes on global food production, primarily through the modification of the prevailing water regimes caused by climatic changes. Many predictions have been made (mostly of a catastrophic nature) on how global food production is going to suffer due to changing patterns of rainfall and temperature, and the attendant problems facing water control and management. There is no question that if the existing rainfall and temperature patterns change during the forthcoming decades, there could be important implications for agricultural production and water management, at global as well as regional and national levels, depending on the rates, magnitudes and spatial distributions of such changes.

## 20.2 Climatic fluctuations and water management

While climatic fluctuations affect water management in a variety of ways, it should be clearly noted from the outset that climate is one of many factors (albeit an important one) that has major impacts on the availability and use of water resources. Among other important factors that affect water management are population levels and densities; extent and level of economic activities, especially in terms of agricultural and industrial development; living standards; and efficiency of water use. These issues are generally closely interrelated.

The most important implications of climatic fluctuations as they relate to water management are in terms of water availability from both surface and groundwater sources; drought and flood management (including efficient operation and safety of reservoirs); choice of proper cropping patterns to ensure crop-water requirements are met reliably; proper functioning of agricultural and urban–rural drainage systems; maintenance of water quality of rivers, lakes, canal systems and aquifers; and management of vulnerable, low-lying areas, especially in the deltaic and coastal regions.

A fundamental aspect of design and operation of any water resources system is to ensure reliable availability of water for various purposes by mitigating the impacts of droughts and floods. Water planners and hydrologists have always considered annual, seasonal, and sometimes daily variations in precipitation as well as river discharges. Dams and reservoirs are designed to explicitly overcome the impacts due to high and low river flows. Generally, design criteria for dams and operating rules for reservoirs are based on past climatological data in terms of rainfall and runoff. It is implicitly assumed that the past experiences and climatological patterns hold the key to the future events. This assumption may not be correct, since current analyses indicate that some of the present hydrological techniques and water management methodologies may not be adequate to successfully deal with the extent of climatic fluctuations witnessed in the past or those that can be expected within the designed life periods of existing or proposed hydraulic structures.

If the world experiences warming due to increases in the concentrations of carbon dioxide and other greenhouse gases, as the majority of the scientists are predicting at present, water availability and use patterns would most certainly change. The rates of such changes would differ from place to place as well as over time. For example, on a global basis, nearly 74% of water used at present is for agricultural purposes. If global warming takes place, irrigation water requirements are likely to increase due to higher evapotranspiration losses. Land use-patterns (including cropping patterns) would change, over a period of time, either by direct policy changes to make more efficient use of scarce water resources due to socio-economic pressure or by a series of *ad-hoc* individual decisions as a direct response to changing climatic regimes. Both will have water-use implications. However, at our present state of knowledge, estimates of the potential climatic changes can, at best, be considered to be informed guesses. We cannot go any further than that because of lack of knowledge and reliable data in many areas. This hinders our understanding of numerous interrelated phenomena and feedback systems which define the various climatic patterns. The current signals of climatic fluctuations are neither strong enough nor observed over a sufficiently long period of time to predict confidently the onset of any climatic change with any reliable degree of scientific accuracy. In spite of such uncertainties, however, many climatologists have already 'confidently' predicted that the global climatic regimes have already started to change due to global warming caused by the greenhouse gases.

It should be further noted that in the area of agricultural production it is not enough to consider changes in annual average rainfall and temperature. Distribution of rainfall and temperature over the cropping period and the beginning and the end of the rainy seasons are crucial factors which determine the yields for rainfed agriculture, which is still the dominant form of agriculture in large areas of the world. Currently, the possible changes in rainfall and temperature patterns in specific regions, even on an annual basis, cannot be predicted with any degree of certainty. Accordingly, it is even more hazardous to forecast inter- and intra-annual changes in rainfall and temperature patterns in specific agricultural project areas. Predictions of changes of onset and end of rainy periods are simply not possible at our existing scientific state of the art. Under these conditions, estimation of changes in crop production due to climatic changes are still too uncertain and vague, and thus cannot be considered for explicit incorporation for any specific planning and management of water and agricultural development projects.

## 20.3 Climatic forecasting

In spite of recent advances in climatology, climatic forecasting is still a very complex and difficult process. Increasing attention must be paid to improve not only the existing monitoring and forecasting techniques but also the facilities available for such activities in almost all developing countries. Without a functional and efficient monitoring and forecasting system, rational planning, design and operation of water resources systems is most difficult (if not impossible) under the best of circumstances.

The global circulation models (GCMs) available at present are still far from infallible. Among many other shortcomings which seriously constrain their actual use for planning and management purposes are better representation of clouds and more reliable modelling of oceans and the way they interact with the atmosphere. Equally very limited data are available from oceans which cover a major part of the world. Available temperature distribution data are primarily based on observations on land surfaces, where often higher temperatures could be noted due to urbanization and other related human activities. Thus, available global surface temperatures are highly skewed in the favour of land surfaces. These shortcomings seriously affect the reliability of predictions of the current generation of GCMs.

The preliminary nature of GCMs can be illustrated by the fact that in late 1988 the global warming forecast by one of the GCMs in about 50 years' time, when concentration of carbon dioxide is expected to double, was around 5.5°C. By changing the way the clouds were treated within the model, the recent prediction by the same model is that the average global temperature would increase by 2–3°C during the same period. Regional changes, as indicated by the model, however, show similar patterns. From a policy point of view, the difference between an average warming of 5.5°C and 2°C could mean devastation or manageable disruption of agricultural production and/or land submergence in coastal areas in many parts of the world. Similarly, economic impacts of the two scenarios are not even close; cost implications and thus investments required to ameliorate potential impacts would be at completely different levels of scale.

The problem is further complicated by the fact that all the models also do not necessarily forecast consistent patterns and/or magnitudes of the potential regional climatic changes. For example, some models predict intensification of monsoon rains over south Asia, which could contribute to increased magnitude of annual floods with consequent devastation and very heavy economic losses. In contrast, others show weakening of monsoon rains, which could mean droughts and crop failures. Policy measures that need to be taken, even at such a macro-scale, to alleviate these two conditions are of greatly different types. Because of such uncertainties, policy makers in general have been reluctant to make heavy investments necessary in advance in anticipation of such potential catastrophic events due to climatic changes.

Even if the predictions made by the global circulation models on such macro-level regional climatic changes were accurate, they are of minor interest at present for water management in terms of planning, design and operation of individual projects. This is because the current generation of GCMs cannot provide usable climatic information on specific river basins, sub-basins or agricultural project areas at a scale that can be used successfully by planners. Equally, general information such as average annual rainfall and average annual runoffs in an area are simply inadequate for planning and management of water resources projects. Hydrological characteristics of water projects must consider and incorporate short-time-scale events, which are extreme in nature such as floods or 10-day low flows for design, safety and operation of dams and reservoirs. Such data are simply not available at present with any degree of accuracy from any of the existing GCMs.

## 20.4 Dilemma of water planners and managers

A major dilemma facing nearly all water planners and managers of developing countries at present is the urgency of developing and managing water projects on the basis of inadequate and/or unreliable data. With the continually increasing pressure resulting from higher population growth levels and demand for better standards of living, developing countries are now trying very hard to increase their agricultural production so that they can become increasingly self-reliant in food availability and simultaneously generate more energy from indigenous sources. Water development is an essential requirement to ensure that a reliable supply of water is available for increasing areas under agriculture and to produce higher crop yields. Since agriculture is an intensive user of water, large-scale water developments are often necessary. The same water resources developed for irrigation also generally produce hydroelectric power. Thus there is often considerable social, economic and political pressure for immediate execution of water projects, irrespective of the quality and extent of hydrometereological data available. Developing countries often do not have the luxury of waiting until reliable data over a reasonable length of time are available before planning and execution of large-scale water development projects can be initiated.

Since water projects often have to be planned and managed on the basis of inadequate and/or unreliable data, a common technique used in recent years has been to generate synthetic streamflow based on whatever short period of data is available. This practice has been so widespread in recent years that it has been generally considered to be a fully acceptable standard procedure. Very few hydrologists or research workers have questioned the efficiency of its use. The main research emphasis has been, almost exclusively, to fine tune the techniques available for synthetic streamflow generation by increasingly complex mathematical manipulations and the use of more powerful computers.

Currently, hydrologists and water resources planners, at least in developing countries, generally consider themselves fortunate if they have 10–15 years of reliable and continuous streamflow data. It is now not uncommon to find that large-scale water projects are being planned on the basis of less than 10 years of reliable hydrometeorological data. On information currently available, and the experience of the recent past, one must question the efficiency and appropriateness of many such designs, and the desirability of continuing the practice of generating synthetic streamflow on the basis of rather short periods of data.

Analyses of climatic fluctuations observed during the past five or more decades in many countries indicate that the concept of generating synthetic streamflow on the basis of short periods of data is flawed, to the extent that its continued use to plan large-scale and high-investment water projects can be seriously questioned. It is becoming increasingly evident that complex mathematical manipulations are no real substitutes for better information if the short period of data on which such manipulations fundamentally depend are unrepresentative of the long-term pattern. The probability that the short period of data available is really representative of the long-term pattern cannot be very high. It is thus imperative that hydrologists and other water resources experts reconsider their research focus so that a new methodology can be urgently developed, which will enable us to handle the river flow fluctuations more effectively within the planning and management framework.

The problem of climatic fluctuations appears to be more complex than the water resources profession has generally realized so far. The magnitude of this problem may vary from country to country, and conceptually it is not going to be an easy problem to resolve. For example, Figure 20.1 shows the annualized Nile flows at Aswan, Egypt, as a percentage of long-term mean covering nearly 120 years. If, as an illustrative example, it is assumed that a major water development project had to be designed in 1900 based on the previous 30 years of data, and in 1950 on the preceding 50 years of data, or in 1990 on the previous 30 years' data, the project designed on the identical river for the same location would most probably have had significant design differences. This means that the benefits and costs of a water project for the identical location could have been very different, depending on the periods of data available to, and used by, the planners.

Similar types of problems would have occurred for the White Nile (Figure 20.2) at Mogren and the Blue Nile at Khartoum (Figure 20.3). The

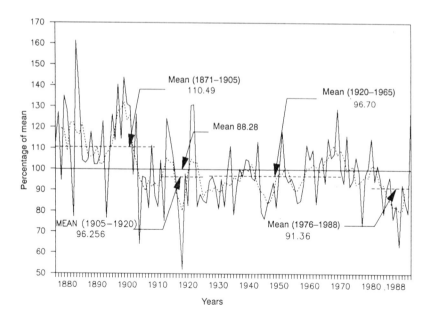

**Figure 20.1**   *Annualized Nile flow at Aswan*

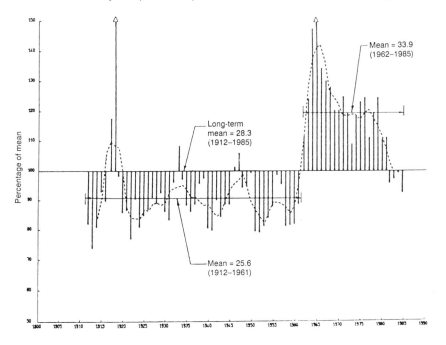

**Figure 20.2** *Naturalized annual flows: White Nile at Mogren*

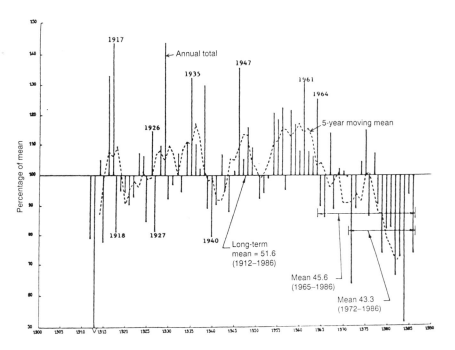

**Figure 20.3** *Naturalized annual flows: Blue Nile at Khartoum*

issue can be seen very clearly in the case of the White Nile at Mogren, where the 73-year river flow could be easily divided into two distinct periods, 1912–1961 and 1962–1985. The mean annual flow during the period 1912–1961 was 25.6 billion m³, but if the period 1962–1985 is considered, it was 33.9 billion m³, which is almost 33% higher than the earlier period. Thus, hypothetically, a water development project for the White Nile at Mogren designed in 1962 on the basis of earlier data would have had very different characteristics, if it had been designed in 1986 on the basis of 1962–1985 data.

Hydrologists and climatologists have often assumed that 30 years of continuous observations can reliably define the statistical parameters of long-term rainfall or runoff patterns. A quick analysis of Figures 20.1–20.4 will clearly indicate that this hypothesis can be seriously questioned in terms of its accuracy. Depending on which 30-year period is selected, the main statistical parameters could be quite different when compared to another 30-year period. The issue is even more complex if shorter periods of observations (say, 10–15 years) are considered for design of water resources projects.

It is possible that the complexity of the problems could vary from one river basin to another, or even from one part of a river system to another for major rivers. It could also vary from country to country, as shown in Figure 20.4 for rainfall patterns for four different Sahelian countries: Burkina Faso, Chad, Mali and Senegal (after Todorov, 1985). Again, depending on which 30-year period is selected, one could obtain some very different results.

Rainfall patterns is an essential consideration for water management for rainfed agriculture. Both cropping patterns and yields depend on the extent and distribution of rainfall during the cropping period. If a cropping pattern was decided in an agricultural area on the basis of an earlier regime, which was comparatively more wet, and the same pattern is continued during a lower rainfall regime, clearly crop yields would suffer. This, in turn, would reduce farm incomes, and under severe conditions, threaten the lives of farmers and their livestock.

Of the four Sahelian countries considered in Figure 20.4 it can be noted that the decline in rainfall was most significant for Senegal, followed by Mali and Burkina Faso, but it was also evident in Niger. A comparison between the earlier isohyets of mean annual rainfall for Senegal with those on the basis of rainfall for the 16-year period of 1968–1983 indicates that the earlier isohyets have to be moved southward by 113 km to reflect the latter pattern (Todorov, 1985). If this period is extended to 30 years, the earlier isohyets have to be moved south by some 61 km. Such changes, of course, have significant implications for water control and management for agricultural production. These are shown in Figure 20.5 (Todorov, 1985).

In addition to the complexities referred to above on how to handle methodologically 'normal' climatic fluctuations within the context of water resources development and management in an efficient manner, another issue further complicates this situation. This is to what extent human

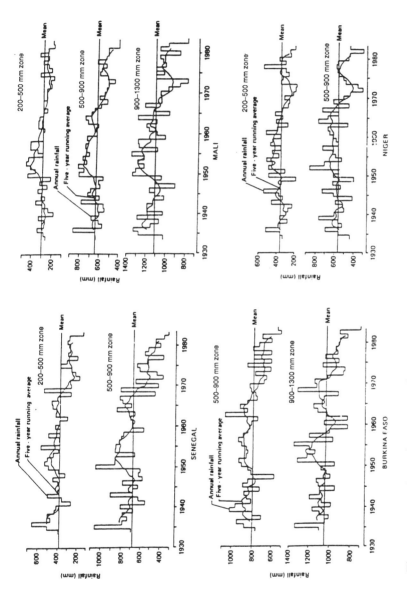

**Figure 20.4** *Annual amounts of rainfall, five-year running average and mean annual rainfall for the 50 years 1934–1983 in Senegal, Burkina Faso, Mali and Niger*

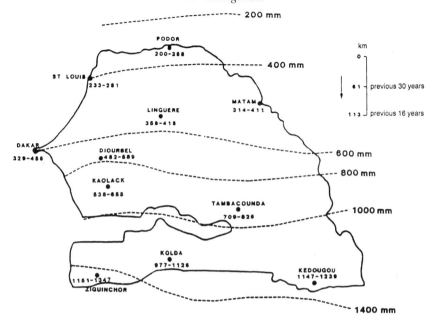

**Figure 20.5** *Isohyets of the mean annual rainfall in Senegal and mean annual amounts for the previous 16 and 30 years for selected stations. Also shown are the distances by which the isohyets should be moved southwards for the more recent averaging periods*

activities such as land-use changes and increased production of greenhouse gases may be changing the climate on a micro- and/or macro-scale. The problem becomes even more difficult if one attempts to separate what percentages of climatic changes are induced by human activities and what are 'normal' climatic fluctuations which are inherent in any climatic regime. While many hypotheses have been put forward in recent years, it has to be admitted that at our present state of knowledge it is simply not possible to answer such important questions with any degree of accuracy. Furthermore, it is highly unlikely – in spite of accelerated research in this area – that this question can be answered satisfactorily during the remaining part of the twentieth century.

Thus, water planners and managers in developing countries are under considerable pressure to provide urgently reliable water control in order to ensure that existing food production and hydropower generation systems could not only be maintained but also be increased to provide a better standard of living for an expanding population base. Simultaneously, on the other hand, the planners are confronted with fundamental methodological problems of developing and managing water projects on the basis of inadequate and sometimes unreliable data, which often makes water planning and management a risky task under the best of circumstances. Furthermore, present experience indicates that some of the recently used so-called sophisticated methodologies used in water management are, in many cases, not as reliable as their proponents have advocated them to be.

During recent years, an important research focus in most parts of the world has been an attempt to predict climatic changes that may occur in the future (say, in 20–60 years' time). This concentration of research emphasis primarily on potential future climatic changes has, unfortunately, appeared to have diverted the attention of hydrologists, climatologists and other water professionals from the urgent necessity of how to cope efficiently with climatic fluctuations, which have already been clearly observed in many river basins during the past several decades. Such fluctuations are not merely a part of a hypothesis: they are a reality. This is an area that must receive priority attention in the near future.

## 20.5 Conclusions

Climatic fluctuations have always been considered to be an inherent part of any water resources planning process. On the basis of recent experience in many parts of the world it is now clear that the existing planning processes and hydrological methodologies need to be substantially improved, at least for certain areas where such fluctuations have been considerable, so that both water management and agricultural development processes could be made optimal on a long-term sustainable basis.

Much concern has been expressed, especially during the past five years, on the prospect of climatic changes induced by global warming due to increasing concentrations of carbon dioxide and other greenhouse gases. While the threat of global warming has to be taken seriously, it is argued that, at our present state of knowledge, not enough detailed information and data are available on potential climatic changes which would enable water resources planners and managers to incorporate such changes within individual water projects in any specific manner. The type of information available at present is primarily of a macro nature, which even if it is considered to be reliable can, at best, be incorporated into the planning process in a very broad way. The problem is made further complicated since, even for such macro-scale information, there is no consensus among scientists on the magnitude, spatial distribution and time scale of occurrence of such changes.

While climatic change is a long-term potential threat, there is no question that proper consideration of climatic fluctuations within the water resources management process is a real, major, current problem. Yet, paradoxically, most of the present research efforts have been directed to the implications of potential climatic changes to water management. Very limited work is being undertaken on how to deal effectively with climatic fluctuations, which can already be identified as a serious problem. Hydrologists, climatologists and other water resources professionals must redress this balance.

# References

Biswas, A. K. (1986) 'Land use in Africa', *Land Use Policy*, **3**, No. 4, 247–259.

Todorov, A. V. (1985) 'Sahel: The changing regimes and the "Normals" used for assessment', *Journal of Climate and Applied Meteorology*, **24**, No. 2, 93–107.

# 21 LARGE-SCALE PROJECTS TO COPE WITH CLIMATIC FLUCTUATIONS

S. Shalash
and
Mahmoud A. Abu-Zeid
*Water Research Center, Cairo, Egypt*

## 21.1 Introduction

Climatological changes are as old as the earth. They are even considered as contributing factors to the equilibrium of the universe. In the Holy *Koran*, for example, two phenomena are cited: one of them is the 'Tofan' that drowned the whole earth except Noah's Ark. The second is the occurrence of seven rainy years over the Nile Valley followed by a 7-year drought period and Joseph's wisdom that could prevent famine catastrophic in Egypt. Thus, climatological changes are as old as history, but their resulting effects are now causing disasters. Hence the interest of scientists whose research is in finding solutions to avoid harmful effects on the environment and human life.

Since the beginning of this century human activities have developed at a fast rate following a rapid increase in population. Such activities, aimed at construction and prosperity, have often resulted in devastation. Man praised everything new, trying all new technologists towards progress. He oriented his efforts on earth, in the deep oceans, and in outer space. His urge to master the world misled him, and he did not give enough consideration to the final results of his activities. Floods and droughts, earthquakes and volcanoes became more frequent. Disasters resulted partly from ignorance and partly from negligence.

In Africa, for example, human activities were stimulated in the mid-century by rapid population growth. Forests were removed, dams were constructed and urbanization expanded. Meanwhile, scientists were recording climatological changes and attributed some of those changes to human activities. For instance, the last twenty years have witnessed droughts and famine in various African countries causing the death of millions. Scientists have referred to such phenomena as regional interferences. But when drought and famine spread over most of the African coastal countries, focus was diverted towards the world climatic cycle and scientists searched for other causes. All countries of the Nile Valley were subject to the effects of drought: Ethiopia, the Sudan, Uganda, Kenya, Zaire, Rwanda, and Burundi, i.e. all except Egypt and North Sudan, in spite of Egypt's location at the mouth of the Nile and hence, its greater exposure to drought (Figure 21.1).

Egypt escaped catastrophe due to the construction of the High Aswan Dam (HAD) – an overyear storage reservoir – and could thus overcome the effects of severe climatic fluctuations. Along with the HAD, other factors helped in the situation, such as the exceptional rise in water level in Lake Victoria, which maintained the high releases of the White Nile and which was due to the abundant 1961–1963 rains. Since then, rains have kept above average. High flows of the White Nile compensated the low releases of the Blue Nile which has been negatively affected by the African drought, attributed by scientists to climatological changes and not solely to regional causes.

The drought phenomenon over the Nile Valley shows that there is a decline in the annual average of natural flow at Aswan. In such cases the decline is shown as the rate of annual flow. The period of high flow at the close of the last century is the decrease of flow by the mid-1960s compared to the 5-year average considered as relatively more recent and reliable than previous periods of low flow. Similar phenomena were observed in Middle African rivers, such as the Senegal, Niger, Zaire and Zambezi, all showing the same but more serious features as the Nile.

There is no doubting the gravity and continuity of the drought phenomenon. The question is whether it is the result of permanent or semipermanent changes in the climatic cycle; and to what extent climatological changes should be considered while making future plans for water development and the methods for operating the current system in order to cope with foreseen changes. The usual method is to rely on statistical facts in estimation of water yield, whereas the two factors of variability and steady state should be based on a physical explanation.

Three factors should be taken into consideration regarding the Nile Basin (or any similar river basin):

1. With continuing drought or overflood, statistics increase so as to affect the estimation of releases.
2. The extension of climatological research gave results that shed light and explained phenomena from a strictly physical point of view.
3. Whether climatic changes are related to natural changes, to human activities, or to both, one should rely on facts rather than on global or regional causes.

This chapter illustrates the method of using the High Aswan Dam to cope with anticipated climatological changes. However, it has now been proved that the global temperature has slightly increased over recent years, and this phenomenon must be taken into consideration when estimating water consumption in agriculture.

**Figure 21.1**

## 21.2 Analysis of meteorological and hydrological records of the Nile Basin during the past century

Meteorological and hydrological records of the Nile Basin started in 1868, and have been maintained for hydrometeorological parameters. There are even older level records of the Nile, such as the Cairo nilometers, which date back for more than a thousand years and which have recorded several climatological changes that affected the Nile in that period. It is known that there are two main sources of Nile runoff:

1. The equatorial lakes and the White Nile from the south;
2. The Ethiopian rivers from the east.

Two main hydrological stations of the Nile Valley were selected to illustrate the historical series over 100 years.

Entebbe Station, which is located on the east shore of Lake Victoria, maintained records from 1896 to 1982. This gauge has recorded the effect of a climatological phenomenon. Rains over the lakes during the period 1961–1963 raised their level by 3 m above average. This effect of high lake levels continued for about 20 years (Figure 21.2 and Table 21.1).

**Table 21.1**   Level averages of Entebbe Gauge Station

| Period | Average level (m) |
| --- | --- |
| 1896–1982 | 10.87 |
| 1896–1960 | 10.57 |
| 1960–1982 | 11.79 |

It was found that the average level had risen since 1961 by about 1 m, as shown in Table 21.2. The average level of the period 1960–1982 was higher than that of 1896–1960. Also, human activities on the lake's outlet were increased by the construction of the Owen Falls Dam in 1956. Therefore, significant changes in normal lake levels may be related to climatic changes, human activities, or both. However, both affected positively the natural river flow for a long period.

The Aswan Gauging Station was selected as the site for computation of the total river yield, disregarding the considerable losses of river flow (i.e. the net river yield for the previous 15 years). Statistical analysis for the Nile's natural flow (Figure 21.3 and Table 21.2) can be as follows. Knowing that the average annual natural flow of the Nile Basin for the period 1871–1988 is 88 milliard $m^3$, it is observed that the 10-year average of the periods before 1900 is more than 84 milliard $m^3/yr$, around 100.850 milliard $m^3/yr$. As for the periods after 1900, 10-year averages alternate around 84 milliard $m^3/yr$, except for the period 1961–1971. It is also noticeable that the

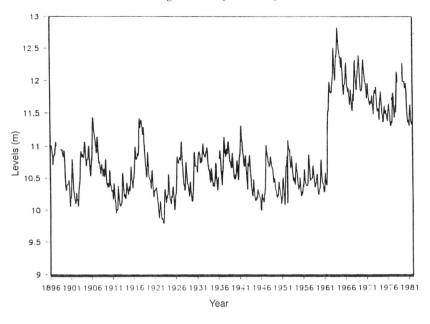

**Figure 21.2** *Entebbe gauge readings (monthly average)*

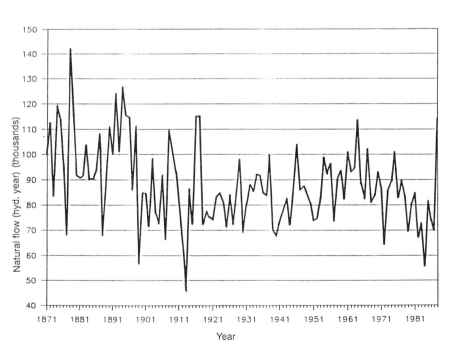

**Figure 21.3** *Natural annual flow at Aswan (milliard m³) from 1871–1872 to 1988–1989*

**Table 21.2**   Natural flow averages

| Period | Averages (milliard m³/yr) |
|---|---|
| 1871–1881 | 104 624 |
| 1881–1891 | 94 160 |
| 1891–1901 | 102 196 |
| 1901–1911 | 86 836 |
| 1911–1921 | 81 022 |
| 1921–1931 | 80 566 |
| 1931–1941 | 84 428 |
| 1941–1951 | 83 593 |
| 1951–1961 | 85 845 |
| 1961–1971 | 93 361 |
| 1971–1981 | 83 380 |
| 1981–1988 | 77 676 |

period 1981–1988 is one of very low annual average flow on the Nile Basin. There is no definite trend for the natural flow of the Nile Basin but it varies around 84 milliard m³/yr.

### 21.2.1 Overyear storage theory

The optimal long-range capacity of the High Aswan Dam has been computed by means of the overyear storage formula that gives the best results on the basis of previous available records of the flow at Aswan. This equation and its computation have been carried out by the late H. E. Hurst. It has also been computed on the basis of all previously analysed similar phenomena, taking into consideration the site's available storage capacity.

The basic equation is as follows:

$$\log \frac{R}{\sigma} = K \log \frac{N}{2} \tag{21.1}$$

where

$R$ = maximum or minimum accumulated storage, or both,
$\sigma$ = standard deviation,
$N$ = statistical variable parameter.

With the application of this equation to the case of the High Aswan Dam and given the following information, namely that (1) the average annual yield in a period of 88 years from 1870 to 1957 is 92 milliard m³; (2) the standard annual deviation is 18 milliard m³; and (3) the most probable value of $K$ is 0.72, it is found that the capacity necessary for storage in the High Dam is 300 milliard m³.

However, any reservoir to be constructed on the Nile will fail to hold such a capacity. Thus, to attain a storage within the limits of the capacity available at the actual site, a below-average discharge would need to be released. The required storage capacity can be computed using

$$\log \frac{S}{R} = 0.08 - 1.05 \frac{(M - D)}{\sigma} \tag{21.2}$$

where

$D$ = average,
$M$ = the deviation below the average,
$S$ = the available storage capacity,
$R$ = the storage capacity calculated in equation (21.1)

A guaranteed fixed discharge which would be below the average yield was estimated to be 84 milliard m³/yr, representing the average of the Nile's discharge in the present century until 1958. Thus, by substituting this amount in equation (21.2), the necessary live storage capacity is found to be 85.2 milliard m³. This means that the capacity set for live storage in the High Aswan Dam, estimated at 90 milliard m³, guarantees the release of an average discharge estimated at 84 milliard m³/yr for a period of 100 years. If one considers that the average losses of overyear storage amount to about 10 milliard m³/yr, then the net Nile water yield which the High Aswan Dam guarantees is 74 milliard m³/yr. This would be shared by Egypt and the Sudan as follows:

55.5 milliard m³ for Egypt
18.5 milliard m³ for the Sudan.

In order to safeguard the function of HAD for 500 years it is estimated that the dead storage capacity should 31 milliard m³. In addition to safeguarding the dam walls during high flood it is estimated that flood storage should be 43 milliard m³. Thus the total capacity of HAD is 164 milliard m³ at high flood level (182 m above MSL).

## 21.3 High Aswan Dam – Egypt's multipurpose project

It is known that the High Aswan Dam was constructed to fulfil multi-purpose objectives which can be summarized as follows:

1. To store in its reservoir all floodwater that might otherwise be lost to the sea;
2. To protect Egypt during high-flood years;
3. To protect Egypt during low-water years (drought);
4. To guarantee Egypt and the Sudan shares of water in low-water years;
5. To produce power;
6. To assist Egypt in converting basin irrigation land (1 million feddans) into perennial irrigation land;
7. To assist Egypt in increasing existing cultivated land (1 million feddans or more);
8. To increase the cultivated area from 350 000 feddans to 1 million feddans of rice crop.

A study was carried out in two phases. The first phase is the period of filling the reservoir to reach its highest designed levels, beginning in 1967. As regards the first period, we have records of the releases and levels of the Nile from 1869 to the present. Analysis of records divided this lapse of time

**Table 21.3**

| Years | U/S levels at the end of filling period | U/S levels at the end of releasing period | Total average of U/S level through the whole year |
|---|---|---|---|
| 1870–1880 | 179.70 | 174.80 | 177.25 |
| 1880–1890 | 179.70 | 174.80 | 177.25 |
| 1890–1900 | 179.50 | 174.50 | 177.00 |
| 1900–1910 | 169.35 | 164.54 | 166.95 |
| 1910–1920 | 170.22 | 164.85 | 167.53 |
| 1920–1930 | 166.24 | 161.12 | 163.68 |
| 1930–1940 | 169.62 | 164.87 | 167.25 |
| 1940–1950 | 168.92 | 163.93 | 166.38 |
| 1950–1960 | 175.68 | 171.15 | 173.39 |
| Average | | | |
| 1870–1900 | 179.63 | 174.70 | 177.16 |
| 1900–1960 | 170.00 | 165.10 | 167.55 |

into two distinct periods: from 1869 to 1899 and the second from 1900 to 1968. During the first period, levels and releases related to average yields were rather high. Average annual yield was 101 milliard m³.

For the period 1900–1960 the average yield was around 84 milliard m³/yr, which guarantees agricultural and power generation water requirements. Tentative experimental balances on the High Aswan Dam assumed the recurrence of all the recorded years and it was found that u/s average levels were as in Table 21.3.

From the above it follows that prior to the first period, the dam's levels were not less than 174.50 m; average levels over the whole year were not less than 177 m (Figure 21.4); thus agricultural and power generation requirements were guaranteed.

In the event of a sequence of years similar to the period 1900–1960 with an average yield of 84 milliard m³/yr, it appears that the u/s average level decreases to 163.68 m, with an average of about 167.55 m over the whole period (Figure 21.4).

From another analysis for the years 1900–1960 it appeared that the average level over the year is 95% of the period, i.e. not less than 162 m,

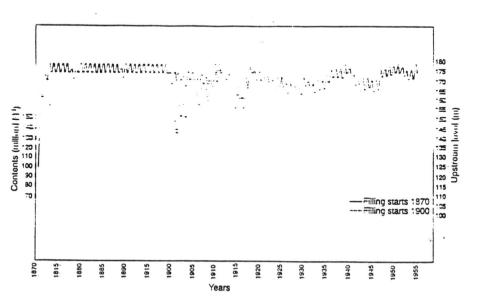

**Figure 21.4** *HAD reservoir management for a period similar to 1870–1955. Note: capacity and surface area of reservoir computed from air survey maps*

thus providing 100% of agricultural requirements, the dam's power generation being 95%. Tentative experimental balances over the dam during the filling period after its completion were based on two assumptions: first, the recurrence of a sequence of the years 1947–1960 representing the actual period of high flow (above average) and, second, recurrence of a sequence of the years 1921–1934 representing an actual period of low flow (above average). The above balances showed maximum and minimum u/s levels during the period 1967–1980, as shown in Table 21.4 and Figure 21.5.

**Table 21.4**

| Years | Sequential high flow | | Sequential low flow | |
|---|---|---|---|---|
| | Lowest level at end of July | Highest level at end of Dec. | Lowest level at end of July | Highest level at end of Dec. |
| 1967 | 138.00 | 155.10 | 130.50 | 149.40 |
| 1968 | 149.20 | 161.70 | 137.00 | 155.70 |
| 1969 | 156.20 | 161.70 | 147.00 | 155.70 |
| 1970 | 169.40 | 169.00 | 154.00 | 163.30 |
| 1971 | 163.00 | 168.20 | 156.20 | 161.00 |
| 1972 | 162.20 | 167.40 | 153.90 | 162.70 |
| 1973 | 161.10 | 186.60 | 156.20 | 161.00 |
| 1974 | 163.00 | 174.10 | 154.20 | 160.80 |
| 1975 | 169.50 | 175.60 | 155.60 | 167.10 |
| 1976 | 171.90 | 178.40 | 161.70 | 165.C0 |
| 1977 | 174.90 | 177.60 | 157.10 | 163.60 |
| 1978 | 172.60 | 179.30 | 156.80 | 164.70 |
| 1979 | 174.90 | 181.40 | 159.00 | 165.10 |
| 1980 | 175.00 | 178.60 | 160.00 | 168.00 |

From study and analysis of the Nile records for the period 1869–1960 it appears that the percentage of frequency of the river discharges for the annual average in the recorded sequence in the above period is as shown in Table 21.5. The expected upstream levels have been assumed in three cases:

1. The first represents average yields;
2. The second represents the case when the flow is 5% of the years, i.e. in 5% of the period the probable flow is 122.00 milliard $m^3/yr$ or more, beginning in the first year of the filling period, with a flow similar to that of the same year and in the sequence of the years up to 1980. Following probability and frequency theories, the gradual decrease of flow reached 100.28 milliard $m^3/yr$ in 1980.
3. The third case represents the occurrence of the flow in 95% of the years, i.e. for 95% of this period the flow is 64.6 milliard $m^3/yr$ and, following

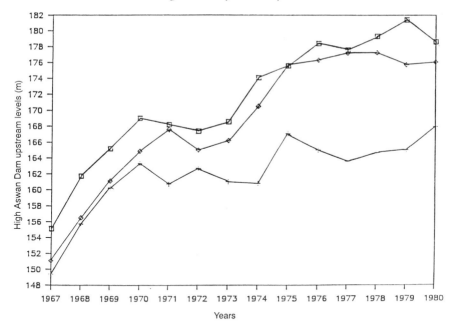

**Figure 21.5** *Forecast upstream levels at 31 December (from High Aswan Dam Study, 1967)*

**Table 21.5**

| | *Percentage of discharge frequency* | | | | | | | |
|---|---|---|---|---|---|---|---|---|
| | 5 | 50 | 60 | 70 | 80 | 90 | 95 | 99 |
| Discharges in milliard m$^3$/yr | 122.0 | 90 | 85.5 | 81.80 | 76.4 | 69.0 | 64.6 | 55.5 |

probability and frequency theories, it reached 82.07 milliard m$^3$/yr in 1980.

Based on the above tentative calculations, expected upstream levels were estimated as shown in Table 21.6.

## 21.4 High Aswan Dam and climatological changes

To show how the dam has coped with wide climatic changes over the Nile Basin from 1968/1969 to 1988/1989, the following must be mentioned:

**Table 21.6**

| Date | Expected upstream levels | | |
|------|--------------------------|--|--|
|      | *When 95% freq. low years* | *Average years* | *When 5% freq. high years* |
| 1 Jan. 1968 | 148.6 | 152.0 | 152.0 |
| 1 Jan. 1969 | 153.4 | 159.5 | 164.5 |
| 1 Jan. 1970 | 156.9 | 164.3 | 171.4 |
| 1 Jan. 1971 | 159.5 | 167.8 | 176.0 |
| 1 Jan. 1972 | 161.7 | 170.5 | 179.4 |
| 1 Jan. 1973 | 163.4 | 172.6 | 180.0 |
| 1 Jan. 1974 | 164.8 | 174.3 | 180.0 |
| 1 Jan. 1975 | 165.8 | 175.6 | 180.0 |
| 1 Jan. 1976 | 166.4 | 176.5 | 180.0 |
| 1 Jan. 1977 | 167.0 | 177.3 | 180.0 |
| 1 Jan. 1978 | 167.6 | 178.0 | 180.0 |
| 1 Jan. 1979 | 168.2 | 178.7 | 180.0 |
| 1 Jan. 1980 | 168.7 | 179.1 | 180.0 |

1. The construction period of the dam lasted for 10 years (1959–1969).
2. The upstream coffer dam of the main wall dam created a part of the reservoir.
3. The high level of Lake Victoria contributed towards extra Nile flow, estimated to be 9 milliard $m^3$ within the period.
4. Few high-water years occurred (1964, 1967) during the construction period which enabled the dead storage part of the dam to be filled (31.6 milliard $m^3$).
5. Also, there were few high-water years (1975, 1976) during the period.

On the other hand, there were few negative incidents:

1. Released discharges downstream from the dam exceeded Egypt's share (55.5 milliard $m^3$ for many years).
2. Several low-water years occurred (African drought from 1973/1974 to 1987/1988).
3. The reservoir water balance of the period 1968/1969–1988/1989 is shown in Table 21.7.

If the dam's releases had been restricted to 53 milliard $m^3$/yr during 1968–1988 the live storage would have increased by 10 milliard $m^3$.

**Table 21.7**

| Years | Nat. flow | Dongola | Contents | Aswan rel. | Losses |
|-------|-----------|---------|----------|------------|--------|
| 1968–1969 | 80.92 | 71.34 | 39.43 | 54.08 | 7.91 |
| 1969–1970 | 83.68 | 71.02 | 45.30 | 55.72 | 8.22 |
| 1970–1971 | 93.33 | 83.00 | 60.60 | 55.50 | 11.06 |
| 1971–1972 | 86.74 | 82.11 | 69.22 | 56.39 | 11.72 |
| 1972–1973 | 64.64 | 50.96 | 56.26 | 56.29 | 8.26 |
| 1973–1974 | 85.68 | 76.83 | 66.98 | 56.29 | 10.90 |
| 1974–1975 | 89.03 | 82.87 | 80.06 | 55.78 | 12.40 |
| 1975–1976 | 111.19 | 102.70 | 108.42 | 53.23 | 15.70 |
| 1976–1977 | 83.21 | 67.40 | 105.14 | 56.11 | 14.20 |
| 1977–1978 | 89.88 | 76.39 | 108.99 | 61.75 | 15.60 |
| 1978–1979 | 83.68 | 72.89 | 111.25 | 59.73 | 15.40 |
| 1979–1980 | 69.33 | 56.11 | 103.21 | 55.75 | 12.10 |
| 1980–1981 | 80.43 | 66.64 | 102.49 | 56.61 | 13.70 |
| 1981–1982 | 85.02 | 68.93 | 99.06 | 59.00 | 14.20 |
| 1982–1983 | 67.23 | 50.95 | 80.92 | 61.23 | 11.20 |
| 1983–1984 | 73.09 | 58.69 | 72.94 | 58.53 | 11.02 |
| 1984–1985 | 55.71 | 40.89 | 51.49 | 56.40 | 7.24 |
| 1985–1986 | 81.83 | 64.57 | 53.70 | 55.52 | 9.80 |
| 1986–1987 | 74.53 | 55.66 | 47.19 | 55.19 | 9.05 |
| 1987–1988 | 69.72 | 50.14 | 41.05 | 52.86 | 7.10 |
| 1988–1989 | 114.28 | | 75.92 | 53.24 | 14.3 |

## 21.5 Overview of flow management

At the beginning of 1988 the upstream level of the dam dropped to one that caused much concern to those responsible. The reservoir level by the end of the water year 1988 (31 July) was expected to be 151.7 m above MSL, which is slightly higher than the elevation at which hydroelectricity cannot be generated (147 m). Recent years have seen low inflow. Annual inflows to the dam's lake in the decade 1976–1985 averaged 76.8 milliard m³, while a long-term (1900–1985) average is 84.1 milliard m³/yr. Consequently, there is debate on how the dam's lake waters should be managed and whether smaller amounts should be released for crops and other uses. The Ministry of Public Works and Water Resources (MPWWR) controls the water releases and partially controls the cropping pattern. Thus, policy decisions can be made regarding cropping patterns and release levels. In turn, policy decision alternatives need to be evaluated in terms of their costs, benefits, risk and distribution implications.

As it was not known by the end of water year 1988 what would be the future climatic changes and their effects on the Nile runoff yield, the Water Planning Group of the MPWWR and the World Bank Panel made a study for a possible similar situation during the next 20 years. The object of the study was to investigate many different options on the dam's operating policy for use by decision makers and answering the following questions:

1. What are the near-term lake level and energy production consequences of releasing different amounts of the dam's lake flow?
2. If various lake level goals are adopted, what levels of releases and energy production will occur?

These questions can be addressed using hydrological, energy and agricultural sector modelling and can be solved with a lake simulation model (WMP).

The dam simulation allowed the development of two types of information. First, the consequences of choosing constant release policies on the lake level were examined to address the question of lake and energy production levels (question 1). Second, the consequences of following alternative lake operating goals on releases and energy production were studied (question 2).

Exploration of the constant release policies starting from a 31 July 1988 lake level of 151.7 m, the 31 July 1989 ending-year lake level using the lake model results in the summary data in Table 21.8. These results simulate what happens to the 31 July 1989 lake level under a particular release policy if the observed inflows in the years 1968–1987 were characteristic of the 1988/1989 flow probabilities starting from a 151.7 m water level in the dam.

Adoption of a 52.5 milliard $m^3$ release policy would yield a lake level increase (a July 1989 level above 151.7 m) in 13 of the 20 flow years, with an average level of 154.3, but in seven of these flow years the lake falls below 151.7, with one of the 20 years falling below 143 and four below 147, with power generation ceasing at least for the latter part of the flow year.

Adoption of 50.5 milliard $m^3$ would still preserve the 31 July 1988 elevation or, better, 14 years out of 20, but would yield a level no lower than 144 m. However, within two years the levels would fall below 147 m, the critical power-generating level. In these six years, levels of 144 m, 147 m, two years of 149 m and two years of 151 m occur. Preventing the lake level from falling below the 31 July 151.7 m level in all years under a constant release policy requires releases of 33.5 milliard $m^3/yr$ or less.

Several conclusions may be drawn from these data under the assumption that the past 20 years is representative of future inflows. First, the results indicate that there is a significant probability of falling lake levels if the current practice of fixed releases is pursued. On the other hand, there is more than a 50% chance that lake levels will increase even under the highest level of releases. It appears that, on average, lake level increase by about 0.3 m per milliard that constant releases are restricted.

**Table 21.8** Frequency of HAD 31 July 1989 water levels under varying releases with inflow for 20 simulated flow years (initial level 151.7 m)

| Levels | Release policy | | | | | | | |
|---|---|---|---|---|---|---|---|---|
| | 55.5 | 53.5 | 52.5 | 51.5 | 50.5 | 49.5 | 47.5 | 45.5 |
| $d < 142$ | 1 | 0 | 0 | 0 | 0 | 0 | 0 | 0 |
| $142 < d < 143$ | 0 | 1 | 1 | 0 | 0 | 0 | 0 | 0 |
| $143 < d < 144$ | 0 | 0 | 0 | 1 | 1 | 0 | 0 | 0 |
| $144 < d < 145$ | 1 | 0 | 0 | 0 | 0 | 1 | 1 | 0 |
| $145 < d < 146$ | 0 | 1 | 1 | 1 | 0 | 0 | 0 | 1 |
| $146 < d < 147$ | 2 | 1 | 0 | 0 | 1 | 1 | 0 | 0 |
| $147 < d < 148$ | 0 | 1 | 2 | 2 | 0 | 0 | 1 | 0 |
| $148 < d < 149$ | 2 | 1 | 0 | 0 | 2 | 2 | 0 | 1 |
| $149 < d < 150$ | 1 | 1 | 2 | 2 | 0 | 0 | 2 | 0 |
| $150 < d < 151$ | 0 | 1 | 1 | 0 | 2 | 2 | 0 | 2 |
| $151 < d < 152$ | 1 | 0 | 0 | 1 | 1 | 1 | 2 | 1 |
| $152 < d < 153$ | 2 | 1 | 1 | 0 | 0 | 0 | 1 | 1 |
| $153 < d < 154$ | 1 | 3 | 2 | 2 | 1 | 1 | 0 | 1 |
| $154 < d < 155$ | 3 | 2 | 1 | 2 | 3 | 2 | 1 | 0 |
| $155 < d < 156$ | 2 | 1 | 3 | 3 | 2 | 2 | 3 | 2 |
| $156 < d < 157$ | 0 | 2 | 2 | 1 | 1 | 2 | 2 | 2 |
| $157 < d < 158$ | 3 | 0 | 0 | 1 | 2 | 2 | 1 | 3 |
| $158 < d < 159$ | 0 | 3 | 3 | 3 | 1 | 0 | 2 | 2 |
| $159 < d < 160$ | 0 | 0 | 0 | 0 | 2 | 3 | 2 | 0 |
| $160 < d < 161$ | 0 | 0 | 0 | 0 | 0 | 0 | 1 | 3 |
| $161 < d < 162$ | 0 | 0 | 0 | 0 | 0 | 0 | 0 | 0 |
| $162 < d < 163$ | 1 | 0 | 0 | 0 | 0 | 0 | 0 | 0 |
| $163 < d < 164$ | 0 | 1 | 1 | 1 | 0 | 0 | 0 | 0 |
| $164 < d < 165$ | 0 | 0 | 0 | 0 | 1 | 1 | 1 | 0 |
| $165 < d < 166$ | 0 | 0 | 0 | 0 | 0 | 0 | 0 | 1 |
| $d > 166$ | 0 | 0 | 0 | 0 | 0 | 0 | 0 | 0 |
| Average | 153.2 | 153.9 | 154.3 | 154.6 | 155.0 | 155.3 | 156.0 | 156.6 |
| Std dev. | 5.2 | 5.1 | 5.0 | 5.0 | 4.9 | 4.9 | 4.7 | 4.6 |

Source: WMP Report

Turning now to energy, restrictions in releases lead to diminished energy production. Each milliard reduction decreases energy production of about between 876 000 and 1 138 000 MGh.

These data permit derivation of flexible strategies designed to maintain a minimum lake level as required to answer question 2. A number of such strategies are presented in Table 21.9. Note, for example, that maintenance

**Table 21.9**   Incidence of releases required to maintain various minimum lake levels (starting from 151.7 m)[a]

| Release (milliard m³/yr) | Minimum lake level to be maintained (m) | | | | | | | | | |
|---|---|---|---|---|---|---|---|---|---|---|
| | 147 | 148 | 149 | 150 | 151 | 152 | 153 | 154 | 155 | 156 |
| 55.5 | 18 | 16 | 16 | 14 | 13 | 13 | 12 | 10 | 9 | 6 |
| 54.5 | 0 | 0 | 0 | 0 | 0 | 0 | 1 | 1 | 0 | 0 |
| 53.5 | 0 | 1 | 0 | 1 | 1 | 0 | 0 | 1 | 0 | 1 |
| 52.5 | 0 | 1 | 0 | 1 | 0 | 0 | 0 | 0 | 1 | 2 |
| 51.5 | 0 | 0 | 0 | 0 | 0 | 1 | 0 | 1 | 1 | 0 |
| 50.5 | 1 | 0 | 2 | 0 | 2 | 0 | 0 | 0 | 1 | 0 |
| 49.5 | 0 | 0 | 0 | 0 | 0 | 0 | 0 | 0 | 0 | 1 |
| 48.5 | 0 | 1 | 0 | 2 | 0 | 1 | 1 | 0 | 1 | 1 |
| 47.5 | 0 | 0 | 0 | 0 | 0 | 1 | 0 | 0 | 0 | 1 |
| 46.5 | 0 | 0 | 1 | 0 | 0 | 0 | 0 | 0 | 0 | 0 |
| 45.5 | 0 | 0 | 0 | 0 | 2 | 0 | 1 | 1 | 0 | 1 |
| 44.5 | 1 | 0 | 0 | 0 | 0 | 0 | 1 | 0 | 0 | 0 |
| 43.5 | 0 | 0 | 0 | 1 | 0 | 1 | 0 | 0 | 0 | 0 |
| 42.5 | 0 | 1 | 0 | 0 | 0 | 1 | 0 | 2 | 1 | 0 |
| 41.5 | 0 | 0 | 0 | 0 | 1 | 0 | 0 | 0 | 0 | 0 |
| 40.5 | 0 | 0 | 1 | 0 | 0 | 0 | 2 | 0 | 0 | 0 |
| 39.5 | 0 | 0 | 0 | 0 | 0 | 0 | 0 | 0 | 2 | 1 |
| 38.5 | 0 | 0 | 0 | 0 | 0 | 1 | 0 | 0 | 0 | 0 |
| 37.5 | 0 | 0 | 0 | 1 | 0 | 0 | 0 | 2 | 0 | 1 |
| 36.5 | 0 | 0 | 0 | 0 | 0 | 0 | 0 | 0 | 0 | 1 |
| 35.5 | 0 | 0 | 0 | 0 | 1 | 0 | 1 | 0 | 0 | 0 |
| 34.5 | 0 | 0 | 0 | 0 | 0 | 0 | 0 | 0 | 2 | 0 |
| 33.5 | 0 | 0 | 0 | 0 | 0 | 1 | 0 | 0 | 0 | 0 |
| 32.5 | 0 | 0 | 0 | 0 | 0 | 0 | 0 | 1 | 0 | 1 |
| 31.5 | 0 | 0 | 0 | 0 | 0 | 0 | 0 | 0 | 0 | 1 |
| 30.5 | 0 | 0 | 0 | 0 | 0 | 0 | 1 | 1 | 2 | 2 |

[a]These data give the number of years a particular release policy needs to be followed in order that the specified lake level or greater is maintained

Source: WMP Report

of a 152 m minimum lake level yields a policy wherein 55.5 milliard m³ are released in 13 of the 20 years, while in the other years releases of 51.5, 48.5, 43.5, 42.5, 38.5, and 33.5 milliard m³ would be necessary. A policy of rebuilding to the 154 m level would permit 55.5 milliard m³ releases in 10 of the 20 years, with other releases as identified in Table 21.9 spanning as low as 1 year with 33.5 milliard m³ or less. Note the higher the lake level goal, the less energy produced in 1988/1989 and the more variable energy production becomes (as measured by the standard error).

Several conclusions may be drawn relative to question 2. The lake may be operated so that certain ending-year levels are maintained, but this requires variable releases and generates variable energy production results. A high target level causes potentially very low levels of releases. Increased energy and lake level variability is encountered, the higher the lake level target.

## 21.6 Conclusion

A long-term capacity reservoir gives water management managers a variety of options in order to avoid any harmful effects on the country's economy due to climatic changes. Seasonal storage dams cannot protect the country from a famine caused by a drought year, as happened in Egypt during the famine year 1913/1914, although Egypt's population at that time did not exceed 15 million.

# 22 A METHODOLOGY FOR MULTI-RESERVOIR MANAGEMENT UNDER CLIMATIC CHANGE CONDITIONS

Kenneth M. Strzepek
*University of Colorado, USA*
and
Juan B. Valdes
*Texas A&M University, USA*

## 22.1 Introduction

This research was conducted to develop a methodology to examine the impacts of climatic change on surface runoff and the resulting impact on the yield of the water management systems within a drainage basin. In this chapter climatic change is defined solely as an increase in annual average temperature with no change in the annual average precipitation. This assumption does not hold in all situations but is adequate for the scope of the study being conducted.

## 22.2 Modelling the climate

In order to analyse the impact of climatic change on a water resource system one must be able to provide some estimate of what the climatic environment will be on a drainage basin scale. Previous and current attempts at modelling global climate change have been made by producing spatial and temporal averages of hydrometeorologic parameters on a large scale. The processes of the hydrologic cycle take place on a much smaller spatial scale and much smaller time scales. Therefore the problem of scale must be addressed. To date, there have been no standard procedures for going from GCM-generated to spatially and temporally varying hydrometeorological parameters.

The time scale of reservoir water yield studies is generally on a month step. However, hydrologic modelling of the continuous water balance of a drainage basin is optimally performed on a daily time step. Therefore climatic change must be modelled on a daily time scale.

The hydrometeorological parameters needed for hydrologic modelling are daily precipitation and daily average temperature. To provide these data, a program that develops synthetic daily weather time series was used. The program WGEN was developed by the US Agricultural Research Service (Richardson and Wright, 1984). This analyses historical time series of maximum and minimum temperature, precipitation, and solar radiation, and calculates stochastic parameters of the time series using Fourier analysis. Based on the Fourier parameter, it will generate synthetic daily time series of maximum and minimum temperature, precipitation, and solar radiation, preserving the statistical properties of the historical time series. These parameters are annual and monthly means, standard deviation and correlation between temperature and precipitation.

With WGEN we can model a climatic change. Since we have defined climatic change as an increase in mean annual temperature, we are able to simply modify the Fourier parameters to include a proposed increase in the mean annual maximum and minimum temperature. We will not modify the precipitation, and in our case we ignore solar radiation. WGEN then generates synthetic time series of daily values that reflect a climatic change.

## 22.3 Modelling the runoff process

Now that a time series of daily data exists that reflects climatic change at a point and on the time scale necessary for hydrologic analysis, we must have a hydrologic model that is appropriate for analysing the impact of this climatic change on the runoff characteristics. Since climatic change was defined as an increase in temperature, the appropriate hydrologic model must be able to model the effect of temperature on the runoff process.

As much of the hydrologic analysis being carried out or proposed for climate change is on basins with little or no runoff modelling and/or little or no streamflow data, the hydrologic model chosen should not be data-intensive. Otherwise all the resources of the analysis will go to hydrologic modelling, instead of impact and management analysis.

In addition, for this analysis the model must be appropriate for mountainous and alpine conditions. Such a model (RRM) was developed for the high alpine lakes of the Salzkammergut of Austria (Fedra, 1989) and its structure is illustrated in Figure 22.1. What is important to note is that not only is the snow/rain distribution and snowmelt a function of temperature, but evaporation and, most importantly, evapotranspiration is modelled as a temperature-dependent process. The model requires very few input parameters, instead assuming default values based on high alpine climate and vegetation for some parameters and calculating others based on the input and default values. The input parameters are basin area, percentage of land covered by forest, pasture, and agricultural land, length of main

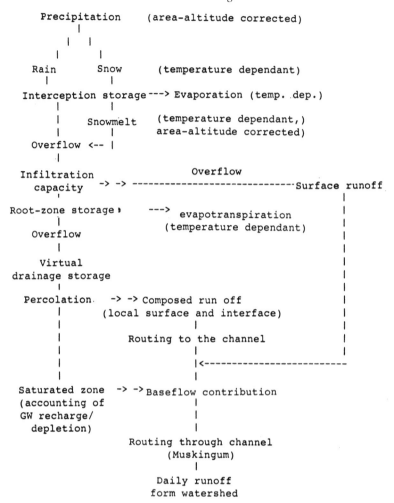

**Figure 22.1** *Rainfall–runoff module for heterogeneous watersheds/sub-watersheds. This estimates daily runoff from daily weather reports (input) and uses daily precipitation totals and average air temperature only, as well as watershed characteristics (parameters) (Interactive Version 7, Dr K. Fedra, Weiszes Kreuzg. 109, A-2340 Moedling, Austria). The program operates on a daily time step and requires daily precipitation (mm) and average air temperature (°C) as input. It also requires observed flow values (m³/s) for calibration and/or performance evaluation. All data must be representative of the lower end of the catchment and altitude corrections are made internally*

channel, length of all major branches, altitude difference in basin and the distribution of the altitude difference into percentage bands.

RRM operates on a daily time step but has the option to write the runoff results to a file as either daily values or monthly summations of the daily values for that month. The model reads the precipitation and temperature data and simulates the daily streamflow at the mouth of the basin. By using the time series from WGEN that reflects a climate change, RRM will produce a time series of streamflow that reflects the basin's response to the change.

## 22.4 Modelling the river basin system

To analyse the impact of the various streamflow scenarios, a simulation model of river basin operation and management is needed. A monthly time step river basin simulation model (MITSIM 2.0) was used, which has been utilized extensively for river basin planning analysis (Strzepek and Garcia, 1989). The model allows the analysis to model water supply and water demand within a river basin and provides for detailed modelling of reservoirs. MITSIM provides for analysis of system-wide and individual project yields and the reliability of these yields. The model is data driven so that it may readily model any river basin if the physical, engineering and hydrologic data are available.

## 22.5 Global warming impact on water supply in the Lower Gunnison reservoir system

*22.5.1 The river basin*

The Gunnison River lies in central Colorado and flows 260 km from the headwaters in Taylor Park below the Continental Divide to the confluence with the Colorado River near Grand Junction, Colorado. This study focuses on the 160-km Lower Gunnison River from the headwaters to the Gunnison Diversion Tunnel in the Black Canyon which encompasses a drainage basin of 9840 km$^2$. There are four reservoirs in series (Cascading) along this stretch of river: Taylor Park (136 MCM), Blue Mesa (919 MCM), Morrow Point (189 MCM), and Crystal (20 MCM). The total storage on the system is 1264 MCM. Blue Mesa, Morrow Point, and Crystal are part of the Colorado River Storage Project to provide water storage for the Lower Colorado Basin states (California, Arizona, and Nevada), which have a total annual water right of 9250 MCM. In addition, there will be power generation, flood control, and recreation benefits to the Gunnison Basin itself. Taylor Park Dam and the Gunnison Diversion Tunnel are part of the Uncompahgre Irrigation Project, which provides irrigation water to south-western Colorado (USBR, 1985).

*22.5.2 The climate*

There is one representative hydrometeorological station for this part of the basin that is to be modelled, and it is located at the Taylor Park Dam. For this analysis, 20 years of data from 1967 to 1986 were used. The mean annual maximum and minimum temperatures for this time series were 9.7°C and −7.8°C, respectively. The mean annual precipitation was 418 mm.

Time series were generated by WGEN from the historic data to produce the 'historic scenario' time series with mean annual maximum and minimum temperatures of 8.9°C and −7.8°C, respectively. The mean annual precipitation was 405 mm.

Time series were then generated for the '+5 scenario' adding 5°C to the mean maximum and minimum annual temperature to become 13.9°C and −2.8°C, respectively. The mean annual precipitation remained at 418 mm.

The mean annual maximum and minimum temperatures for the '+5 scenario' time series were 14.0°C and −2.7°C, respectively. The mean annual precipitation was 405 mm.

### 22.5.3 The streamflow

The 'historic scenario' time series, when used in RRM with the proper basin data, generated a time series of streamflow with a mean annual volume of 1700 MCM. This translates to a rainfall–runoff coefficient of 0.46. The '+5 scenario' time series when input to RRM with the proper basin data generated a time series of streamflow with a mean annual volume of 950 MCM. This translates to a rainfall–runoff coefficient of 0.25.

In addition to the reduction in flow between the historic and +5 scenarios due to increased evapotranspiration, there is a reduction in annual snowfall in the basin. There is also a one-month shift in the flow peak due to a shift in the main snowmelt runoff from June to May.

### 22.5.4 The reservoir system

A schematic of the Lower Gunnison River Basin system is shown in Figure 22.2. The system was modelled to reflect the cascading operation of the four-reservoir system. The yield from the system is to maintain year-round minimum flows in the Black Canyon and to meet irrigation demands in the basin and downstream. Since irrigation demands are temporally varying, a standard time distribution of water requirements that reflects the cropping pattern and irrigation practices of the Uncompahgre Valley was used. Therefore each annual target yield is distributed into twelve monthly yields as follows: Oct.-9.06; Nov.-3.74; Dec.-2.27; Jan.-2.27; Feb.-2.05; March-3.84; April-11.41; May-13.37; June-12.54; July-13.75; Aug.-13.47; Sept.-12.24.

The storage to mean annual flow ratio for the historic scenario is 0.74 and 1.34 for the +5 scenario. Thus the impact of reservoir storage on system performance will be greater in the +5 scenario even though the annual yield will most probably be lower.

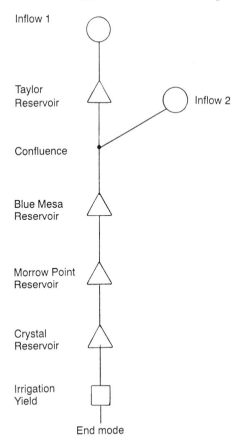

**Figure 22.2** *Schematic of Lower Gunnison river system*

### 22.5.5 Results

Figure 22.3 presents the average annual yield of the reservoirs system plotted against the target yield. It is interesting to note that in both scenarios the actual yield starts to fall below the target yield at approximately the mean annual yield of the natural streamflow without any reservoir storage. Also, the maximum attainable mean annual yield is approximately equal to the sum of the mean annual yield of the natural streamflow and the total storage in the system.

As the target yield increases, the actual mean yield can no longer meet the target with 100% reliability, and a deficit occurs. When this happens it is claimed that the system is less reliable, therefore in addition to a measurement of the system yield, a measure of system reliability must be chosen. Reliability in this work is defined as the ratio of number of times a yield target was completely achieved over the number of attempts to meet the target.

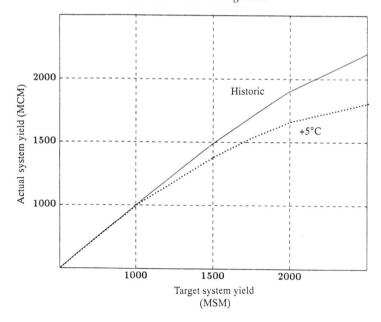

**Figure 22.3**   *Average annual yield of the Gunnison system*

As mentioned above, the annual yield is distributed into monthly yields. Figure 22.4 is a plot of the total monthly reliability for the system under both flow scenarios. In this case, with 20 years of monthly simulation there were 240 attempts to meet the monthly targets. A reliability of 75% means that in 180 of the months the target was achieved.

However, Figure 22.5 is a plot of total annual reliability. This is where a reliability of 75% means that in 15 years out of the 20 the total annual target was not obtained. This is a very strict definition of reliability, because a system may have a small deficit in only one month yet it would be considered that the entire year had a deficit. Note that for the historic scenario in Figure 22.4 the total monthly reliability for a yield of 2000 MCM is approximately 93% while the yearly reliability from Figure 22.5 is 65% and mean annual yield from Figure 22.3 is 95% of the target yield.

## 22.6 Conclusions

This chapter has demonstrated that a methodology for a physical-based climatic impact on water supply can be developed for the drainage basin scale. The case study has illustrated a number of generic observations as well as providing information specific to the Lower Gunnison River Basin. The major conclusions are as follows:

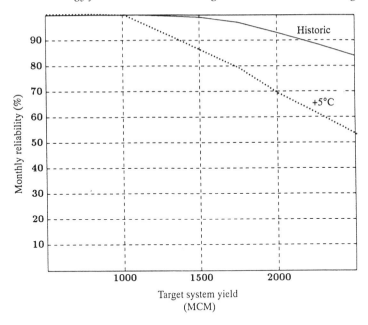

**Figure 22.4** *Total monthly reliability of the Gunnison system*

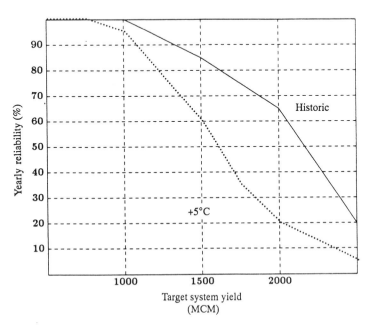

**Figure 22.5** *Yearly reliability of the Gunnison system*

1. The WGEN appears to adequately model the hydrometeorological variables producing time series statistically similar to historical series.
2. The 5°C increase in mean annual temperature significantly reduces runoff in this basin.
3. Reduction in mean annual yields is not linearly related to reduction in runoff.
4. Criteria chosen to establish yield affect the perceived climatic impact.
5. Reliability is impacted greater than mean annual yield in this basin.

# References

Fedra, K. N. (1989) 'RRM: Temperature dependent daily rainfall-runoff model', CADSWES Working paper 20, IIASA, Laxenburg.

Mitchell, A. (1989) 'Greenhouse and climate change', *Review of Geophysics*, **27**(1).

Richardson, C. W. and Wright, D. A. (1984) 'WGEN: A model for generating daily weather variables', US Department of Agriculture, Agriculture Research Service, ARS-8.

Strzepek, K. M. and Garcia, L. A. (1989) 'MITSIM2.0: River basin simulation model', CADSWES Technical Report 1.

US Bureau of Reclamation (1985) *Project Data*, Technical Report.

# 23 CANADIAN PERSPECTIVES ON FUTURE CLIMATIC VARIABILITY

Peter J. Reynolds
*Past President, International Water Resources Association, Nepean, Canada*

## 23.1 Introduction

Climate change presents more than just another environmental problem. A changed climate creates a new framework for the manifestation of other environmental stresses. Acidification of soils and surface water, perhaps a rise in mean sea level of about 1 m in the next century, degradation of forests by air pollution, eutrophication of fresh and salted waters, slower water flows, and soil erosion due to land changes are some of the regional scale problems currently arising from industrial and agricultural development. Stratospheric ozone depletion, occurring rapidly but episodically on a global scale, is increasing the ultraviolet flux to the earth's surface. In the next 40–80 years temperature and precipitation changes due to the enhanced greenhouse effect will exacerbate these environmental problems.

In addition to any difficulties which humans or natural resources may experience in adjusting to the direct consequences of climate change, its intensification of existing pollution problems is an important indirect consequence of a warmer climate. In this chapter we shall focus on existing environmental problems, the possible response by water resources managers, and some strategies for greenhouse defence.

## 23.2 Timing

How much time do we really have? Can we afford to wait until something drastic happens, perhaps 40 or 80 years from now? Timing of measures to manage the undesired effects is the most challenging issue as the regional large-scale measures (fight, adapt or retreat) will have a lead time of 20–40 years before they become effective. After many qualitative studies about climatic effects it is now time to carry out a number of site-specific, quantitative in-depth studies to describe the economic, public health, environmental, social and administrative effects. A preliminary study of the possible impacts of a one metre rise in sea level at Charlottetown, Prince Edward Island, was completed recently (Environment Canada, 1988a). What strategies do we need to manage their effects (e.g. low-lying industrialized coastal zones)? Perhaps such studies should be carried out under the umbrella of a UN organization such as UNEP, and by a global network of scientists and policy makers.

## 23.3 Canadian experience

The experience of Canada shows the following trends consistent with global climate change:

1. Decreasing spring runoff, reducing water supply in the prairies and increasing competition for available reservoir storage;
2. Increasing rainflood frequency, requiring redesign and upgrading of water control structures and flood control dams;
3. Increasing demands for instream and downstream flow releases to protect the ecosystem from further degradation due to reduction of spring flows.

## 23.4 Canadian study highlights

*Estimating Effects of Climatic Change on Agriculture in Saskatchewan, Canada (Environment Canada, 1988b)*

1. This is one of five case studies on cool temperate and cold regions considered in the IIASA/UNEP project to assess the impacts of climatic change and variability on food production.
2. In an extreme drought year, Saskatchewan can expect moisture resources so reduced that the wind erosion potential is doubled, spring wheat production is about 25% of normal and losses to the agricultural economy in 1980 dollars exceeds $1.8 billion and 8000 jobs with a further reduction in other sectors of the economy of $1.6 billion and 17 000 jobs.
3. Occasional extreme 5- or 10-year periods in Saskatchewan can be expected over which biomass dry matter production is reduced by nearly half and spring wheat production by about one-fifth, average annual agricultural losses of about 2600 jobs and $80.6 billion, and a reduction in the provincial GDP ot more than $0.5 billion and 5600 jobs.
4. A shift to a longer-term warm climate with precipitation increases, without major adaptive changes by agriculture, would reduce wheat yields by 16%, causing annual losses to agriculture of over $160 million and 700 person-years. Wind-erosion potential could be reduced, poten-

tial biomass productivity increased, but droughts will become more frequent and severe.

5. A shift to a longer-term warm climate without precipitation increases would cause all impacts to be generally adverse and more intense, particularly as a consequence of increased drought frequency and severity.

*Exploring the Implications of Climatic Change for the Boreal Forest and Forestry Economics of Western Canada (Environment Canada, 1989c)*

1. An extensive literature review was undertaken regarding climatic change impact on tree growth and physiology, direct effects of $CO_2$ enrichment, forest productivity, forest zonation, forest fires, insects and diseases, and forest economics. Many areas requiring further study were identified.
2. Changed climatic resources were calculated and described for several climatic change scenarios based on general circulation model (GCM) results. For example, growing degree-day results indicate increases of 13–48% above normal for the climatic change scenario gridpoints examined.
3. The potential northward shift of the boreal forest was examined using the 600 and 1300 growing degree-day isolines as approximations of the northern and southern boundaries of the boreal forest. The climatic change scenarios indicate considerable shifts in the forest's climate zones as a result of 'greenhouse gas' warming. The potential northward shift in the northern boundary ranges from about 100 to 700 km and the potential northward shift in the southern boundary from about 250 to 900 km.
4. A climatic index of potential biomass productivity and its component factors were used to explore the impacts of the climatic change scenarios on the boreal forest. Examples of potential biomass productivity changes include decreases from normal for southern locations of about 2% to 12% and increases up to about 50% above normal for gridpoints in the northern part of the predominantly forest zone.
5. A multiple linear regression model relating forest productivity to climate was developed and tested. Even with the limited data, there is a significant positive relationship of forest growth rate to both growing season temperatures and moisture availability.
6. A climatic zonation model for the boreal forest was designed and tested. Results of the impacts of one of the climatic change scenarios were calculated and discussed. Preliminary results indicate that, under climate warming, the aspen parkland ecoclimatic zone would extend northward into the western part of the study area and the boreal temperate ecoclimatic zone westward into the eastern part of the study area.
7. A dynamic economic model for analysing the economic benefits of climatic change was identified. The problem of determining the

appropriate objective function and constraints for such a model was addressed and some preliminary qualitative estimates were made. In addition, it was determined that there were important non-market amenities that also had to be taken into account in the objective function. Needs for further research and linkages with the non-economic team members were identified.

8. A systems analysis approach was developed to illustrate and explore the climatic change/forest/economics/policy interactions and an inter-disciplinary integration was applied to the study.

*$CO_2$-Induced Climate Change in Ontario: Interdependencies and Resource Strategies (Environment Canada, 1988d)*

1. Almost all components of the climate system and resource use are affected by $CO_2$-induced climatic warming.
2. The nature and magnitude of these impacts vary considerably across resource sectors.
3. The impacts of $CO_2$-induced climate change are intricately interdependent. Directly affected climate system components and resource uses can, in turn, affect other sectors.
4. Some of these indirect impacts reinforce the effects of $CO_2$-induced climate change; others are mitigative.
5. A variety of resource and socio-economic strategies exist to mitigate the effects of climate change. These include both preventive and adjustment strategies and are based on steady-state assumption, including no changes in the current socio-economic environment and no significant technological changes.
6. While some of these strategies are marginal behavioural modifications, others require significant forward planning (e.g. forest genetics).
7. The timing and pattern of $CO_2$-induced change are important in determining the most likely and desirable strategies.
8. A five-point framework was developed to evaluate strategies and identify research priorities. These criteria are: resource costs; changes in employment and income; geographic distribution of impacts; issues of national sovereignty; and net economic benefit of $CO_2$ control.
9. Research priorities were identified in four broad areas: atmospheric circulation models; effects on the climate system; effects on resource uses; and resource and socio-economic adjustment strategies.
10. Among resource and socio-economic adjustment strategies, research into the forestry sector is of primary importance.

## 23.5 Water resources management response to climate change

The following are possible changes that will be made in response to climate change:

1. *Re-allocation of water resources*: Water resource exploitation has usually been carried out to further specific social goals rather than purely economic ones. Examples are increasing the population of the western US, or industrializing Third World countries such as Ghana or Sri Lanka. Social goals have been reflected in political water allocation decisions or water rights procedures. With increasing competition for a scarcer resource, these decisions must be re-evaluated. The outcome of such a re-evaluation would probably give higher priority to urban and industrial water supply than irrigated agriculture. Thus it is likely that there will be reductions in the amount, and significant reductions in reliability or irrigation supply. Examples of this may be giving priority to the demands of the city of Cairo from the Aswan Dam, or the city of Los Angeles from the Central Valley.

2. *A re-evaluation of water resource economics*: Increasing competition for river flows and reservoir storage increases their value. Because large-scale irrigation schemes are almost always based on political rather than economic rationale, they will become increasingly less attractive to build and maintain. For example, the actual marginal cost of new reservoirs in the USA is of the order of $1/m^3$ of storage; the average water delivery cost is of the order of $0.1/m^3$. Only urban and industrial users can now afford the true costs of water delivery in many semi-arid areas.

3. *Increased incentives for alternative agricultural production*: At present there is little incentive to explore alternatives to large-scale cash-crop irrigation. However, other productive alternatives exist, such as dryland farming, less water-intensive food crops, and water-conservation measures (such as drip irrigation). At the same time, there will be an increased incentive to maximize production in areas where rainfall is plentiful.

4. *Increased incentives for energy conservation*: Decreasing power production and less reliability will increase the effective cost of hydropower, providing an additional incentive for energy-conservation measures.

5. *Increased incentives to institute flood management rather than flood control*: A greater flood frequency and decreased flood control reservoir effectiveness will increase the flood hazard in existing floodplain areas. In many instances additional costs of levee reconstruction or dam spillway modification will not be as economic as floodproofing and floodplain zoning in controlling flood damage.

6. *Integration of ecosystem needs in water resource planning*: In many instances fisheries, estuarine ecosystems and riparian woodlands have direct benefits, supporting the livelihood of many people. Traditionally, these flows have been ignored by water resources developers. Now they must be considered. If flows are reduced below certain limits, the ecosystem will be destroyed.

7. *Redesign of water engineering facilities*: Large-scale investment in hydraulic engineering infrastructures may become obsolete with changing climate, requiring redesign and reconstruction. Examples are: increasing the size of canals, pipelines and pumping plants to accommodate greater variability in runoff, relocation of intake structures due to river

channel changes, and replacing spillways on dams to accommodate larger floods.

In order for policy makers to respond effectively to climate change, the first requisite is to understand it as a resource management question, not a hydraulics engineering or plumbing problem. The efficient, sustainable use of water resources requires the resolution of many competing goals. Approaching this question with a broader perspective will require major rethinking of existing policies.

## 23.6 Conclusion

We are not looking for one solution but a set of solutions. Certainly, the proposal to establish an $18 billion global fund for greenhouse defence is excellent, and more influential people must join in the fight to preserve and sustain our natural resources. In terms of being prepared for action we must start asking ourselves questions as to the state of our preparedness. What shape are our reservoirs in? Do we have an up-to-date inventory on their capacity and present use? Should we plan for more? What state of readiness are our flood-protection works? Where are our weak defences? What are the plans for emergency preparedness?

Finally, the issues of future water supply and demand are central to climate impact assessment, and of particular concern to Canada. The challenge is to consider 'adaptive' management strategies that enhance the resistance of the regional ecosystem, and that would make sense whether the climate changes or not.

## Note

The views expressed in this chapter are solely those of the author and not necessarily of the Government of Canada.

## References

Adams, R. M. and Crocker, T. D. (1989) 'The agricultural economics of environmental change: some lessons from air pollution', *Journal of Environmental Management*, **128**, 295–307.

Environment Canada (1988a) *A Preliminary Study of the Possible Impacts of a One Metre Rise in Sea Level at Charlottetown, Prince Edward Island*, CCD-88-02, 1–2.

Environment Canada (1988b) *Estimating Effects of Climatic Change on Agriculture in Saskatchewan, Canada,* CCD-88-06, 1–9.

Environment Canada (1989c) *Exploring the Implications of Climatic Change for the Boreal Forest and Forestry Economics of Western Canada,* CCD-89-02, 1–18.

Environment Canada (1988d) *$CO_2$-Induced Climatic Change in Ontario: Interdependencies and Resource Strategies,* CCD-88-09, 1–12.

# 24 EFFECT OF THE PREVIOUS DROUGHT EVENT ON NILE RIVER YIELDS

Sarwat H. Fahmy
*Chairman, Egyptian Public Authority for Drainage Projects*

## 24.1 Facts about drought in Africa

1. Drought can be described as a decrease in the quantity of water needed for essential usage caused by climatic conditions in a certain area and occurs in areas of both high and low rainfall. Frequent drought events have recently received increased attention in attempts to develop water resources to cope with the requirements of populations in areas prone to drought.
2. By observing drought events in countries of central and coastal Africa, it appears that the duration periods are between one and up to seven years, with the cycle recurring every 10–11 years.
3. Rainfall levels in south Sudan and Upper Egypt vary from 0 to 1500 mm over the Ethiopian plateau, Kenya and Uganda.
4. Eighty per cent of the total area of the Nile Basin can be considered arid or semi-arid, and cannot tolerate long periods of drought, as its agriculture depends on the floodplains in the lower levels of the wadi. Long droughts during this century occurred in 1913–1914 and 1943–1944. The years 1972–1973 were dry, exacerbating the prevailing droughts in coastal Africa, Ethiopia and the Sudan. Table 24.1 shows the years of drought in Ethiopia, indicating affected areas and severity.
5. The problem received serious attention after recent climatic fluctuations in the region and another drought, and recurrence can be expected in the future. Accordingly, the ability to predict the beginning of a drought event and its duration period is very helpful for the countries likely to be affected. For example, in Australia several studies have been conducted on improving methods of predicting drought, but they have not yet been completed.

*24.1.1 Analysis of data on the nature of the Nile and its yield*

This chapter is concerned with data analysis of the Nile during this century, relating it to drought in the region in the last two decades. After

**Table 24.1** Occurrence of drought and famine in Ethiopia

| Year | Affected area | Severity |
|---|---|---|
| 253–242 BC | Ethiopia | Deduced from the chronology of low Nile levels |
| AD 1066-1072 | Ethiopia and Egypt | Low Nile levels |
| 1252 | Refers to Ethiopia as a whole | Low Nile levels |
| 1272–1273 | Refers to Ethiopia as a whole | Low Nile levels |
| 1274–1275 | Refers to Ethiopia as a whole | Low Nile levels |
| 1435–1436 | Refers to Ethiopia as a whole | Low Nile levels |
| 1454–1468 | Refers to Ethiopia | Low Nile levels |
| 1543–1544 | Refers to Ethiopia | Low Nile levels |
| 1560–1562 | Especially in Harar | For three years following the killing of Emperor Gelaudios there was no rainfall, especially in Harar (Eastern Ethiopia) |
| 1800 | Refers to Ethiopia as a whole | Both men and horses died of famine |
| 1826–1827 | Refers to Ethiopia as a whole | There was great failure of both cotton and grain crops and many cattle died |
| 1829 | Shoa Region | Crop failure occurred and a cattle epidemic followed |
| 1835 | Shoa (Central Ethiopia) and Western Eritrea Region | Many people of Shoa (Central Ethiopia) died following failure of rain. This may be the drought that the people of western Eritrea (north-western Ethiopia) remembered as the Great Starvation, when 'rain disappeared from the earth and famine came over to men and beasts'. As Penkhurst suggests, it would have been a major widespread drought |
| 1836 | Northern provinces, Wolloc, | A holocaust of drought, famine, cattle epidemic and cholera. In |

| Year | Affected area | Severity |
|------|---------------|----------|
| | particularly Lastic region and West Eritrea | south-west Eritrea it was a continuation of the 1835 famine, referred as 'the year of stagnations' |
| 1888–1892 | Ethiopian highlands and lowlands | This was one of the most serious droughts experienced in Ethiopia and was known in Ethiopia history as the 'Kifuken' All the rains failed. The entire period was hot and dry and the effects of drought were magnified by the catastrophic renderpest epidemic (which killed 90% of the cattle) and invasions of locusts, caterpillars and rats. Suicide and cannibalism occurred and wild animals attacked people. About a third of the population perished |
| 1895–1896 | Refers to Ethiopia as a whole | A minor drought occurred this year, due to the failure of the winter and spring rains, yet many people and cattle died |
| 1899–1900 | Refers to Ethiopia as a whole | An unrecorded drought was caused by a fall in the level of Lake Rudolf. The Nile flood was also abnormally low |
| 1913–1914 | Ethiopia (northern part especially) | Very low Nile flood (the lowest since 1695). In the northern part of Ethiopia, the price of grain increased thirtyfold and there was great starvation in Tigrai |
| 1921–1922 | Refers to Ethiopia as a whole | Similar drought to that of 1965–1966. According to the recollection of an English resident in Ethiopia, there was a complete lack of rain from October 1920 to May 1921 |
| 1932–1934 | Refers to Ethiopia as a whole | The level of Lake Rudolf dropped, implying a serious decrease in rainfall in southern Ethiopia. A |

| | | drought was recorded in northern Kenya and in 1934 a relief camp was set up in British Somalia to aid the victims |
|---|---|---|
| 1953 | Wollo and Tigrai regions | Another undocumented drought in Wollo and Tigrai |
| 1957–1958 | Wollo and Tigrai regions | A complete lack of rain in this period, together with outbreaks of locusts, epidemics and famine. The worst year was said to be 1957, which did not have more than ten rainy days. More than 100 000 people died. Plagues of locusts in September and October 1958 had a devastating effect on all the awarajas of Tigrai and probably on neighbouring regions |
| 1964–1965 | In Ethiopia as a whole | A virtually undocumented drought, said to be more widespread in Ethiopia than that of 1973–1975. |
| 1965–1966 | Wollo and Tigrai regions | The failure of the spring and summer rains and the high temperatures that accompanied the drought affected five of the eight awragas of Tigrai and eight of the twelve awragas of Wollo were reported. The number of cattle, pack animals, sheep and goats lost is estimated to be between 287 and 350 for Tigrai alone |
| 1969 | Britrca region | Severe drought affecting 1.7 million people |
| 1971–1978 | Northern, south-eastern and eastern parts of Ethiopia, particularly Tigrai | Complete failure of the spring rains. RRC, in its publication of December 1982 gives the number of dead to be about 200 000 for Tigrai, Wollo and Northern Shoe. Other estimates give 400 000 to 1 million for Tigrai and more than 100 000 for Wollo. 80% of the cattle, 50% of the sheep and 30% of goats perished |

| Year | Affected area | Severity |
|------|---------------|----------|
| 1975–1976 | Wollo and Tigrai regions | The four awragas of western Wollo and parts of Tigrai suffered poor harvests and the main harvest of 1977/1978 was destroyed due to 'unfavourable climatic conditions'. In Wollo alone about 1.2 million people were affected. For both Wollo and Tigrai, estimated figures were 2–3 million people |
| 1978–1979 | Southern Ethiopia | Failure of spring rains resulted in drought |
| 1982 | Northern Ethiopia | Delay of monsoon rain by two months, which could have resulted in great loss of life. The government took immediate and effective steps to aid people, and the effects of drought were minimized |

the construction of the High Aswan Dam, the dam's lake started to fill in 1968, which provided Egypt not only with its annual requirements but also with long-term storage in years of low yield. The lake began to fill gradually until storage reached 134 milliard m³ in 1978. Figure 24.1 shows the storage in the dam's lake since 1968. After 1978 the level of the lake started to gradually decrease, then fell sharply after 1981, as seen in Figure 24.1.

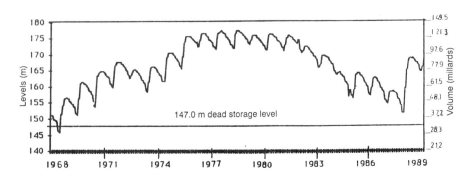

**Figure 24.1**    *Lake Nasser storage, 1968–1989*

## 24.1.2 Available water quantity in the Nile Basin

The drought which has been prevalent over the last two decades has seriously affected several regions in African countries. The question being raised is whether or not this event will continue. The most affected area is coastal Africa, as this originally had high rainfall, and this change in the climatic cycle has led to severe problems. Egypt has not been affected by drought so far due to long-term storage in the High Aswan Dam's lake. Many other aspects have also helped to limit drought in Egypt, the first of which is the high water levels in Lake Victoria caused by heavy rainfall over its catchment in the period 1961–1963. The high water level in the lake has helped to increase the White Nile discharges, compensating for the loss in those of the Blue Nile. Figure 24.2 shows the water levels in Lake Victoria after 1935, and the increase in the years following 1960 is significant.

Figure 24.3 shows the annual discharge of the White Nile at the junction of the Blue and White Niles at El-Mogran in milliard m³, and the normal for the years 1912–1985. Increases above normal after 1962 are considerable. Figure 24.4 shows the annual yield of the Nile at Aswan in comparison to long-term normals, depicting the reduction in yield over recent years. Figure 24.5 shows the annual discharge fluctuations of the Blue Nile at Khartoum, in comparison to normals for the years 1912–1986 and a decrease in discharge since 1964 is obvious. Figure 24.6 shows the same with reference to the Atbara River. The same phenomenon occurred in all rivers of central Africa, but the problem has been worse for the Senegal and Niger rivers than for the Nile.

## 24.1.3 Reasons for drought

Scientific research in this field has given many different reasons for the causes of drought events, which can be summarized as follows:

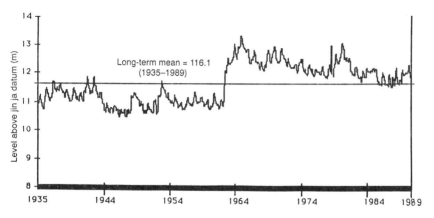

**Figure 24.2**  *Lake Victoria level, 1935–1989*

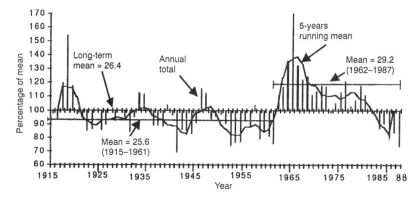

**Figure 24.3**   *Annual flows of the White Nile at Mogren (milliard m³)*

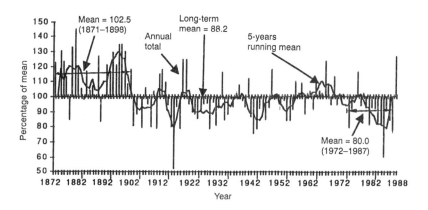

**Figure 24.4**   *Annual flows of the Nile at Aswan (milliard m³)*

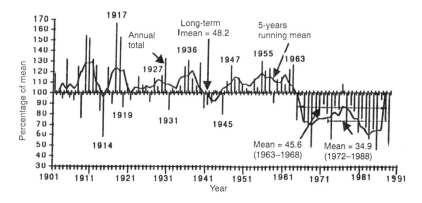

**Figure 24.5**   *Annual flows of the Blue Nile at Khartoum (milliard m³)*

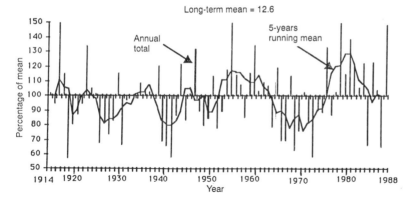

**Figure 24.6** *Annual flows of the Atbara (milliard $m^3$)*

1. Deterioration of the ozone layer as a result of excess carbon dioxide in the atmosphere due to fuel burning. While there was a concentration of carbon dioxide in 1880 of 270 ppm this has increased by 25%, to reach 340 ppm in 1980. If this trend continues, the concentration may reach 500 ppm in the next 50 years.
2. In 1960 there was a noticeable increase in temperature in the lower mid-global section as opposed to a reduction in oceans in the upper mid-section. After 1980 a reduction in temperature occurred, which resulted in an increase in both sections followed by a change in the sea surface temperature. This led to a decrease in rainfall on coastal areas.
3. Continuous encroachment on forests and destruction of vegetation in the biosphere are followed by a decrease in soil moisture.

### 24.1.4 Effects of drought

The previous drought event in countries of coastal and central Africa has shed light on the problem and has attracted attention due to its adverse effects on national economies. It also showed the need for further research to determine the causes and effects of drought, and for planning to reduce the severe effects of future drought events in Africa. As the consequences are severe and long term, this requires governments in drought-prone regions to prepare plans and policies.

In January 1985, 21 African nations faced severe reductions in food, which required the FAO to send aid in the form of nutrition and livestock with a value of $243 million to the countries affected, equivalent to approximately half of the FAO's total external aid ($512 million). However, due to an increase in rainfall levels, the requirements fell from 7 million tons in 1985 to 2.7 million tons in 1986.

The effects of drought are as follows:

1. Famine and its consequences as well as socio-economic effects:
   (a) Migration from affected areas to highly populated regions, causing problems of overpopulation.
   (b) The return of these migrants to their original homes or to new places of shelter after the drought would takes a long time and requires large amounts of money.
   (c) The prevalence of disease and epidemics due to malnutrition and lack of sanitary conditions in refugee camps.
2. Death of livestock and wildlife.
3. Reductions in forests and grazing areas.
4. The drying out of lakes and floodplains and subsequent effects on aquaculture.
5. One of the most severe effects of drought is loosening of the topsoil, which leads to erosion. A UN report states that Ethiopia loses approximately one billion tons of its topsoil every year through such exposure.
6. The continuation of drought for a long period leads to desertification due to encroachment by sand in vegetated areas.

## 24.2 Suggested procedures to counter drought

Drought events in Africa are a natural phenomenon, but over the last two decades their rate has increased in duration and severity, due to the fact that the Nile derives its waters from rainfall over the Ethiopian plateau and equatorial lakes. Therefore droughts in Egypt are affected by the level of rainfall in those two catchments.

The Nile's basin, like those of all rivers, has periods of both high and low yield.. Prediction of drought is both uncertain and inaccurate, and thus procedures for a strategy to control surface and groundwater resources to counter the destructive effects of future droughts and reducing their severity are as follows:

1. Cooperation between countries of the Upper Nile and those downstream by collecting data, recording information, constructing and exchanging information systems;
2. Regulation of Nile flows by building dams and storage facilities at tributaries of rivers, especially in drought-prone areas, to store rainwater in years of extensive rainfall. These may also be used for production of electricity and in deepening of shallow river channels.
3. Securing and protecting river banks against high flooding to decrease the amount of water lost to swamps;
4. Swamps are a major cause of river water waste. When the river narrows, increased discharges lead to its overflowing and thus to the formation of swamps during years of high rainfall. When rainfall levels

drop the water does not return to the main stream due to the fact that the plains are below the river level and as the temperature rises the water is lost through evaporation and evapotranspiration. Generally, in Africa (and in the Nile Basin in particular) water is lost to swamps in great quantities every year. An example of this is the amount of annual water loss in the swampy areas of Bahr El-Gabal (approximately 15 milliard $m^3$, on average). To save this water, the Jonglei Canal Project has been initiated with the objective of conveying excess water in Bahr El-Gabal to the White Nile through the Jonglei Canal outside the swamp areas.

5. Searching for and recording groundwater, setting up groundwater maps, and planning for optimal utilization of groundwater without affecting the storage balance.

6. Expanding hydrological and meteorogical studies of the Nile and its tributaries to cover the regions of the equatorial lakes and the Nile Basin catchments in Ethiopia. A precise analysis of data should be made to develop both surface and groundwater for their optimal utilization. A good example of this cooperation is the hydrometeorological studies of the equatorial lakes (Victoria, Kayoga, Mobuto, Sea Seco).

7. In each country, irrigation methods should be improved and use of modern techniques encouraged (for example, sprinkler or drip irrigation, especially on newly reclaimed land). Training farmers in the use of these modern methods is a necessity.

8. Development and improvement of field irrigation is required for irrigating agricultural land that is difficult to treat by modern methods.

9. Rationalization of municipal and industrial water and use of potable groundwater as a substitute for surface water;

10. Encouragement of scientific study and research in the following fields:
    (a) Prediction of drought over the long term;
    (b) Prediction of rainfall and different runoff ratios contributing to the discharge of each tributary;
    (c) Studies related to water losses from the surfaces of lakes, storage facilities and swamp areas through evaporation and evapotranspiration.

11. Reconsideration of crop patterns and selection of crops with low water consumption as well as increasing the production levels of agricultural land to achieve the maximum income with a smaller water consumption;

12. Alerting people through the media to the water levels and encouraging them to economize on water usage in all areas;

13. Training in modern technology, computer programs and mathematical models, for better communication and exchange of information between all African nations, especially those in the Nile Basin, as well as other countries with similar conditions.

# Part 4

# Case Studies and Reports

# 25

# POTENTIAL IMPACTS OF GLOBAL CLIMATIC EVOLUTION ON THE OPTIMAL SCHEME OF A HYDROELECTRIC COMPLEX: INFLUENCE ON THE NATURAL STREAMFLOWS

Steven Weyman
and
Pierre Bruneau
*Hydro-Québec,*
*Montreal, Canada*

## 25.1 Introduction

The Romaine river basin covers approximately 13 000 km$^2$ and is located on the north shore of the gulf of St Lawrence, halfway between Sept-Iles and Natashquan in the province of Quebec. It flows from its source in the high plains of Labrador from north to south into the gulf of the St Lawrence river (see Figure 25.1).

The projected hydroelectric complex includes four reservoirs and three power plants. The upstreammost reservoir is located on the St Jean river just west of the Romaine river and is used to divert part of the waters from the upper reach towards the Romaine basin.

According to the latest development plan published by Hydro-Québec, the projected sites should be in operation some time after the year 2006. The first objective of this research project is to generate streamflow series for the four projected sites of the Romaine complex based upon future climatic conditions that may prevail during its economic life. The use of a rainfall–runoff model such as the SSARR model (US Army Corps of Engineers, 1987) serves that purpose precisely.

This chapter compares streamflow data for a period between 1938 and 1987 with those computed for the period extending between 2006 and 2055 at gauging station 073802. This station, shown in Figure 25.1, monitors the drainage from a 6550 km$^2$ basin.

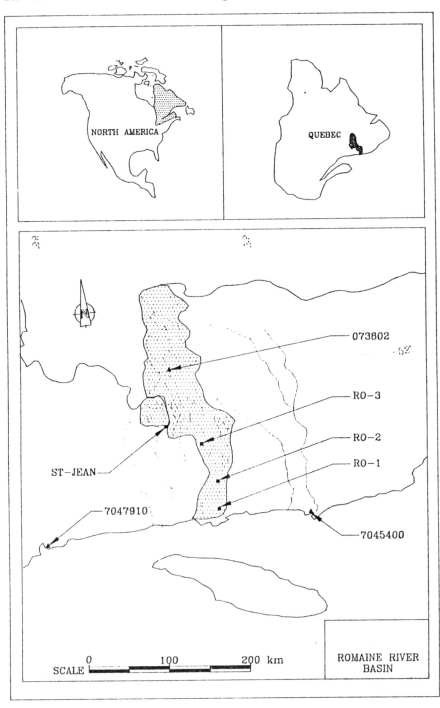

**Figure 25.1**   *Romaine river basin*

## 25.2 Calibration of the SSARR model

The SSARR watershed model was calibrated to simulate the response of the basin to precipitations and temperatures observed at neighbouring meteorological stations. Calibration parameters include the evapotranspiration as a function of temperature, elevation and precipitation intensity and the separation of the runoff into four types of flows with storage times ranging between hours and months. The model also provides features for streamflow routing through channels, lakes and reservoirs. Finally, SSARR Version 8.0, which is used in this chapter, has the capability of computing the accumulation and the melting of snow.

Few meteorological stations exist in the region under study. In order to have a sufficiently long period of record of precipitation and temperature, the observations at stations 7045400 in Natashquan and 7047910 in Sept-Iles were used to calibrate the model. The data from these stations are published by Environment Canada. Their location is also shown in Figure 25.1. Even if both stations are located outside the Romaine basin, the model calibration over the period 1972–1982 yields satisfactory results.

Daily streamflows observed at station 073802 are available for that 10-year period. The accuracy of the simulated discharges depends mainly on the calibration parameters, but the meteorological records may occasionally reflect isolated conditions that do not prevail over the whole region. The calibration process was carried out in such a way that discrepancies are minimized.

## 25.3 Future climatic conditions

Global circulation models (GCMs), surface energy-balance models (SEBMs) and radiative–convective models (RCMs) have been used throughout the world to simulate the climatic conditions that are likely to occur at equilibrium when the tropospheric concentration of carbon dioxide is increased. Schlesinger (1987) reviewed the simulations of five GCMs that give a variety of temperature and precipitation changes for a doubling of the pre-industrial concentration of carbon dioxide. Recent publications identify other trace gases such as methane, nitrous oxide and chlorofluorocarbons that also contribute to enhance the greenhouse effect (Dickinson and Cicerone, 1986; Houghton and Woodwell, 1989; Jager, 1986; Mintzer, 1987).

Table 25.1 shows the global results from five GCM/mixed-layer models (Schlesinger, 1987). It can be seen that the expected global temperature increase ranges between 2.8° and 5.2°C. The time required to reach equilibrium after an abrupt $CO_2$ doubling depends on the mixed-layer depth of the oceans and the model used and varies between 50 and 100 years. Global precipitation may also increase by 7–15% according to these

**Table 25.1**   Changes in the global-mean surface air temperature and precipitation rate simulated with GCM/mixed-layer ocean models for a carbon dioxide doubling (Schlesinger, 1987)

| Model | Temperature increase (°C) | Relative precipitation increase (%) |
|---|---|---|
| GFDL/Wetherald and Manabe (1986) | 4.0 | 8.7 |
| GISS/Hansen *et al.* (1984) | 4.2 | 11.0 |
| NCAR/Washington and Meehl (1984) | 3.5 | 7.1 |
| OSU/Schlesinger and Zhao (1988) | 2.8 | 7.8 |
| UKMO/Wilson and Mitchell (1987) | 5.2 | 15.0 |

figures. The GCM models are three-dimensional numerical ones. The latitudinal distribution of their results all agree, indicating a magnification of the warming at high latitudes compared to the low ones. There is, however, uncertainty in the spatial distribution of precipitation.

A scenario analysis of future use of fossil fuels made by Nordhaus and Yohe (1983) indicates an 80% chance that the carbon dioxide concentration will have doubled from the pre-industrial level by the year 2100. Jager published in 1986 an article based on material presented at the Villach Conference in Austria. In this article it is clearly shown that $CO_2$ doubling may also occur in the middle of the next century. This hypothesis depends mainly on the rate of consumption of fossil fuels and of release of the above-mentioned trace gases. However, there is an upper limit to the temperature rise. Idso (1982) suggested that, unlike the observed greenhouse inferno on the surface of Venus, the mean annual temperature on earth will never exceed an upper limit of 320°K (47°C).

Hoyt (1979) found an empirical relation between sunspots' umbrae/penumbrae cycles and the temperature variation in the Northern Hemisphere. His work reveals that over the period 1880–1970, the global temperature increase due to anthropogenic gases alone was 0.4°C or less. Hansen *et al.* (1981) brought support to Hoyt's theory. They used a one-dimensional radiative–convective model to compute temperature as a function of altitude. Their simulation of past temperatures included the effects of increased carbon dioxide concentration, volcanic aerosols and Hoyt's hypothesis of solar variability. They noted that Hoyt's empirical finding improved the results of their model.

For the purpose of this study, a reasonable scenario of future climate change is necessary. Schlesinger (1987) mentions that discrepancies in GCMs' results come from their validation of the present climate. Because of the uncertainty involved in an accurate prediction of future climate on a regional scale, the meteorological observations at stations 7045400 and 7047910 were examined to provide some clues towards a probable scenario. Station 7045400 gives an almost complete set of temperature and precipitation observations for the period 1915–1987. A complete set of data covers the period 1945–1987 at station 7047910.

A comparison of the mean annual temperatures for these stations shows that they are almost perfectly parallel with only a few tenths of a degree difference. Figure 25.2 shows the mean annual temperature at station 7045400 with a 5-year running mean to filter the noise.

Hoyt's empirical relation on solar variability inspired the use of a similar procedure applied to the annual mean temperature curve. A temperature-forcing function, shown in Figure 25.3, closely fits the running mean curve, leaving only residuals with zero mean and a standard deviation of 0.756. The function has four components:

1. A cosine wave accounting for a short-term cycle;
2. A cosine wave describing a longer-term cycle;
3. An exponential function describing the temperature increase due to anthropogenic production of carbon dioxide;
4. A constant value for the intersect on the temperature axis.

The function parameters were manipulated in an iterative process in order to minimize the sum of squares of the residuals and to bring the mean of the residuals as close to zero as possible. The exponential function was set for a temperature rise of 4.0°C between 1890 and 2100 due to carbon dioxide and an initial temperature difference of 0.4°C for the period 1890–1980 (the iterations stopped with that value increased to 0.5°C). The final form of the equation is:

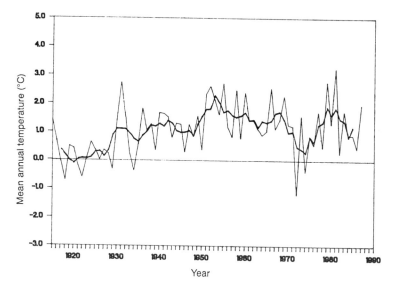

**Figure 25.2** *Observed mean annual temperatures (Natashquan – station 7045400)*

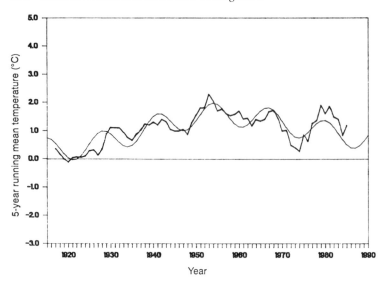

**Figure 25.3**   *Temperature-forcing function (Natashquan – station 7045400)*

$$T(°C) = 0.42 \times \cos [2\pi \times (year - 1954)/13]$$
$$+ 0.54 \times \cos [2\pi \times (year - 1955)/76]$$
$$+ 0.1051 \times \exp [0.0173 \times (year - 1890)]$$
$$+ 0.70 \tag{25.1}$$

The cosine waves of this empirical temperature-forcing function have periods of 13 and 76 years for the short- and the long-term cycle, respectively. This equation is used to calculate the regional temperature trend of the twenty-first century for the Romaine basin.

To obtain the mean annual temperatures for the 2006–2055 period, the residuals calculated in the period 1938–1987 are simply added to the temperatures predicted by equation (25.1). Figure 25.4 shows a plot comparing the temperatures for both series. Then the annual temperature differences relative to the base case are disaggregated into monthly values which, in turn, are added to the daily temperatures. The disaggregation pattern is shown in Figure 25.5. It is derived from the simulation results of the National Center of Atmospheric Research (NCAR) obtained in 1984 by Washington and Meehl. The assumed 4.0°C increase in this scenario also comes from the NCAR model predictions at latitude 52°N.

The regional predictions for precipitation changes are quite uncertain, since they vary from one model to another. The time distributions of precipitation changes are similar at latitude 52°N according to the results of the five GCMs reviewed by Schlesinger (1987). However, the spatial distributions show positive and negative variations for the region under study. Therefore the precipitation values used in the future streamflow

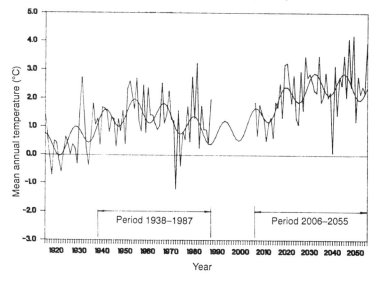

**Figure 25.4** *Hypothetical annual temperature (Natashquan – station 7045400)*

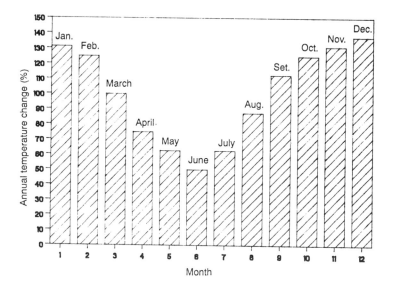

**Figure 25.5** *Monthly temperature increase (source: NCAR, 1984)*

simulations at station 073802 for the period 2006–2055 are assumed to be the same as for 1938–1987.

## 25.4 Simulation results

Two series of simulations were performed with the SSARR model to calculate the daily streamflows at station 073802. The base case covers the period 1938–1987 and uses the observed temperature and precipitation data of stations 7035400 (Natashquan) and 7047910 (Sept-Iles). It should be mentioned here that the meteorological data are missing during the period 1938–1944 at station 7047910. Therefore, only the data of station 7045400 are used to simulate the flows at station 073802 over that period. Since both stations have approximately the same mean annual temperature and the same annual precipitation, the use of both stations aims at averaging their observations to index the meteorological conditions of the basin located between.

The results for the reference case are satisfactory. The interannual discharge obtained for the period 1938–1987 at station 073802 is approximately 136 m$^3$/s, which represents a 9% decrease from that calculated for the observed period 1972–1982. This discrepancy is quite small if one considers the difference in the number of years and the relatively higher hydraulicity of the 1970s.

The simulation of the period 2006–2055 requires the daily temperatures calculated with the procedure described in Section 25.3. In this scenario the precipitation values are the same as those used in the base case.

The interannual temperature increase for the period 2006–2055 is 0.88°C compared with 1938–1987. The algorithm used in the SSARR model to calculate the evaporation is based on a temperature index. The Thornttwaite method is used to establish the relation between temperature and potential evapotranspiration.

Tables 25.2 and 25.3 summarize the magnitudes obtained for evaporation and discharge on a monthly basis. The corresponding interannual evaporation over the 50-year period shows an increase of only 6 mm (2.9%), which translates to a theoretical decrease of 1.2 m$^3$/s for the interannual discharge. On average, the greatest change occurs in April, with a relative increase of 17.5%. In general, due to subfreezing temperatures during the cold months between November and March, the evaporation does not change significantly. It is worth noting that there are actually no evaporation observations in the basin under study to validate the SSARR model's output.

Table 25.3 shows that the simulations for the 2006–2055 period yield an interannual discharge increase of 0.3 m$^3$/s, which contradicts the theoretical decrease expected above. This discrepancy comes from an 80% correction factor that was applied to the precipitation depths observed at subfreezing temperatures in the calibration process: the snow water

293

**Table 25.2** Monthly evapotranspiration in the basin of station 073802 (mm)

*Period 1938–1987*

|  | Jan. | Feb. | March | April | May | June | July | Aug. | Sept. | Oct. | Nov. | Dec. | Annual |
|---|---|---|---|---|---|---|---|---|---|---|---|---|---|
| Mean | 7.1 | 6.1 | 6.5 | 7.2 | 22.3 | 38.7 | 37.3 | 27.2 | 19.6 | 13.7 | 9.7 | 8.6 | 203.8 |
| Minimum | 6.2 | 5.7 | 6.2 | 6.0 | 11.8 | 25.2 | 28.5 | 17.1 | 10.8 | 9.0 | 6.0 | 6.2 | 138.6 |
| Maximum | 10.5 | 8.5 | 9.3 | 12.6 | 36.6 | 51.6 | 53.0 | 34.7 | 29.4 | 19.5 | 15.3 | 12.7 | 293.8 |

*Period 2006–2055*

|  | Jan. | Feb. | March | April | May | June | July | Aug. | Sept. | Oct. | Nov. | Dec. | Annual |
|---|---|---|---|---|---|---|---|---|---|---|---|---|---|
| Mean | 7.5 | 6.2 | 6.8 | 8.4 | 23.9 | 37.9 | 36.0 | 27.9 | 20.9 | 14.8 | 10.1 | 9.1 | 209.6 |
| Minimum | 6.2 | 5.7 | 6.2 | 6.0 | 12.4 | 25.2 | 26.4 | 17.4 | 11.1 | 9.0 | 6.0 | 6.2 | 137.7 |
| Maximum | 12.1 | 10.5 | 12.1 | 14.7 | 36.6 | 53.1 | 51.2 | 35.7 | 28.8 | 21.4 | 16.2 | 13.6 | 305.8 |

*Difference observed for period 2006–2055*

|  | Jan. | Feb. | March | April | May | June | July | Aug. | Sept. | Oct. | Nov. | Dec. | Annual |
|---|---|---|---|---|---|---|---|---|---|---|---|---|---|
| Mean | 0.5 | 0.1 | 0.3 | 1.3 | 1.6 | −0.8 | −1.2 | 0.7 | 1.3 | 1.2 | 0.5 | 0.4 | 5.8 |
| Minimum | 0.0 | 0.0 | 0.0 | 0.0 | 0.6 | 0.0 | −2.2 | 0.3 | 0.3 | 0.0 | 0.0 | 0.0 | −0.9 |
| Maximum | 1.5 | 2.0 | 2.8 | 2.1 | 0.0 | 1.5 | −1.9 | 0.9 | −0.6 | 1.9 | 0.9 | 0.9 | 12.1 |

**Table 25.3** Average monthly discharge at station 073802 (m³/s)

*Period 1938–1987*

| | Jan. | Feb. | March | April | May | June | July | Aug. | Sept. | Oct. | Nov. | Dec. | Annual |
|---|---|---|---|---|---|---|---|---|---|---|---|---|---|
| Mean | 42.26 | 34.77 | 31.48 | 48.56 | 363.49 | 321.78 | 181.74 | 145.82 | 141.96 | 145.97 | 109.21 | 64.45 | 135.96 |
| Minimum | 27.41 | 25.47 | 25.18 | 27.35 | 127.20 | 180.41 | 113.29 | 62.20 | 37.16 | 51.07 | 42.52 | 32.45 | 97.19 |
| Maximum | 89.22 | 79.44 | 69.81 | 220.13 | 666.86 | 647.87 | 287.57 | 248.31 | 231.89 | 271.46 | 260.16 | 157.27 | 180.86 |

*Period 2006–2055*

| | Jan. | Feb. | March | April | May | June | July | Aug. | Sept. | Oct. | Nov. | Dec. | Annual |
|---|---|---|---|---|---|---|---|---|---|---|---|---|---|
| Mean | 47.86 | 38.75 | 36.43 | 74.27 | 363.40 | 281.82 | 171.95 | 141.39 | 139.17 | 145.67 | 117.03 | 76.75 | 136.21 |
| Minimum | 27.41 | 25.66 | 25.79 | 27.64 | 133.48 | 146.25 | 100.85 | 61.97 | 36.45 | 50.77 | 44.02 | 33.10 | 97.05 |
| Maximum | 109.98 | 166.25 | 147.80 | 287.12 | 765.68 | 547.97 | 280.22 | 241.97 | 226.65 | 271.99 | 255.80 | 239.05 | 183.14 |

*Difference observed for period 2006–2055*

| | Jan. | Feb. | March | April | May | June | July | Aug. | Sept. | Oct. | Nov. | Dec. | Annual |
|---|---|---|---|---|---|---|---|---|---|---|---|---|---|
| Mean | 5.59 | 3.98 | 4.96 | 25.72 | -0.09 | -39.96 | -9.79 | -4.43 | -2.79 | -0.31 | 7.82 | 12.29 | 0.25 |
| Minimum | -2.69 | -1.50 | -0.74 | -1.70 | -173.21 | -175.92 | -40.91 | -11.89 | -8.13 | -6.34 | -15.13 | -5.92 | -6.80 |
| Maximum | 46.23 | 86.81 | 77.99 | 179.21 | 210.71 | 27.93 | 8.01 | 4.44 | 2.36 | 22.88 | 60.22 | 90.20 | 9.18 |

equivalent is overestimated in the precipitation data or the snow precipitation observed at stations 7045400 and 7047910 is greater than the actual snowfall over the basin area. The error always shows up during the spring flood. The rating curve of station 073802 may also be inaccurate for high discharges. To prevent this type of error, the 80% correction should be made directly to the snow precipitation data. Nevertheless, the error is negligible in this scenario since the theoretical interannual discharge decreases by only 1%. In fact, the most valuable information that can be extracted from these two cases lies in the time distribution of streamflows. Figures 25.6 and 25.7 show the summary hydrographs with 1%, 50% and 99% probabilities of non-exceedance for the periods 1938–1987 and 2006–2055, respectively. These plots show a 25% decrease of the median spring flood peaks accompanied by a thaw 6 or 7 days earlier. For clarity, the median summary hydrographs are plotted simultaneously in Figure 25.8. On a monthly basis, Table 25.3 reveals a gain in discharge between November and April (the winter season), which is compensated by losses in the remaining months and more substantially in June.

These observations are true for almost all the individual annual hydrographs after the year 2025. Before that time, all hydrographs are almost identical. The average temperature change between the two 50-year series is 0.88°C. Figures 25.9 and 25.10 show sample plots comparing the hydrographs of 1976 (with 2044) and 1977 (with 2045). The mean annual temperature increase between the years 1976 and 2044 is 1.4°C, resulting in a decrease in the spring flood peak of 30%. Note that the flows in autumn increase because of a delay in the advent of winter. This runoff reduces the snow accumulation of winter 2044–2045, leading to a substantial reduction

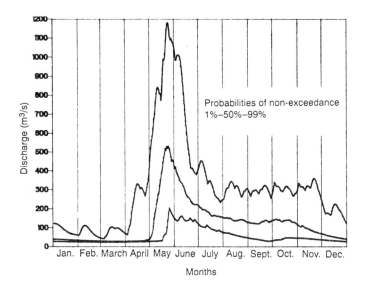

**Figure 25.6** *Summary hydrograph – period 1938–1987 (station 073802)*

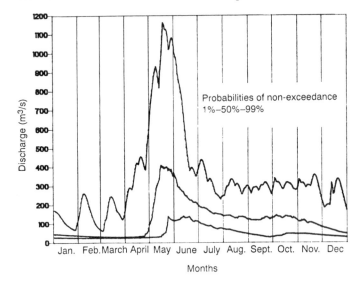

**Figure 25.7**   *Summary hydrograph – period 2006–2055 (station 073802)*

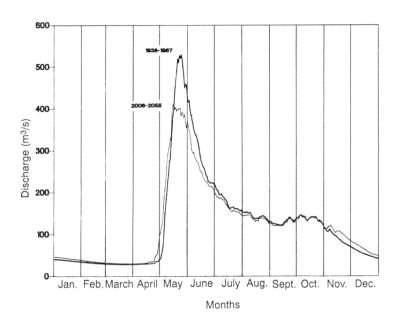

**Figure 25.8**   *Median hydrographs at station 073802 for the periods 1938–1987 and 2006–2055*

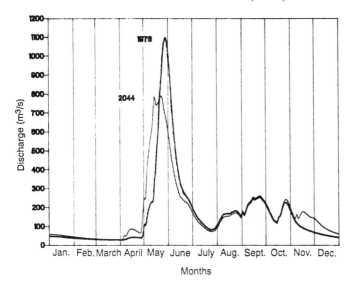

**Figure 25.9** *Hydrographs for the years 1976 and 2044 (station 073802)*

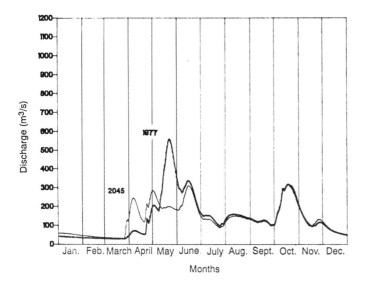

**Figure 25.10** *Hydrographs for the years 1977 and 2045 (station 073802)*

of spring flood volume and peak in the year 2045. The temperature change between the years 1977 and 2045 is only 0.1°C.

## 25.5 Conclusion

This chapter has described an approach to determine the streamflow hydrographs at a streamflow gauging station under a reasonable scenario of a global temperature increase of 4.0°C by the end of the twenty-first century. The hypothesis used for future temperatures is based upon the combination of recent global circulation model results and an empirical determination of the temperature trend observed locally. The precipitation data were not modified due to lack of reliable information.

Simulations performed with the SSARR model on a daily basis show that the interannual discharge at station 073802 is almost unchanged. The main impact of this scenario resides in a gradual modification of the runoff distribution with the increasing number of years. The flows actually observed in winter are increased while the spring floods occur 6 to 7 days earlier, on average, with reduced peaks.

If a more rapid temperature change due to the doubling of carbon dioxide is reached around the year 2050 instead of 2100, more pronounced changes in the runoff distribution should be expected and runoff should decrease substantially. Year-to-year simulation results show such a tendency, with the most severe cases observed in the last years of the period under study.

## References

Dickinson, R. E. and Cicerone, R. J. (1986) 'Future global warming from atmospheric trace gases', Nature, 319, January, No. 6049, 109.

Hansen, J. et al. (1981) 'Climate impact of increasing atmospheric carbon dioxide, Science, 213, August, No. 4511, 957.

Houghton, R. and Woodwell, G. (1989) 'Le réchauffement de la terre', Pour la science, No. 140, June, 22.

Hoyt, D. V. (1979) 'An empirical determination of the heating of the earth by the carbon dioxide greenhouse effect', Nature, 282 (5737), November, 388.

Idso, S. B. (1982) 'Temperature limitation by evaporation in hot climates and the greenhouse effects of water vapor and carbon dioxide', Agricultural Meteorology, Amsterdam, November, 105.

Jager, J. (1986) 'Climatic change: floating new evidence in the $CO_2$ debate', Environment, 28, September, No. 7, 6.

Mintzer, I. M. (1987) A Matter of Degrees: The Potential for Controlling the Greenhouse Effect, World Resources Institute Research Report 5, April.

Nordhaus, W. D. and Yohe, G. W. (1983) 'Future paths of energy and carbon dioxide emissions', *Changing Climate*, National Academy of Sciences, Washington, DC.

Schlesinger, M. E. (1987) 'Model projections of the equilibrium and transient climatic changes induced by increased atmospheric $CO_2$', *Impact of Climatic Variability on Canadian Prairies Symposium*, Edmonton, 9–11 September.

US Army Corps of Engineers (1987) *SSARR Model – Streamflow Synthesis and Reservoir Regulation, User manual*.

# 26 CLIMATIC CHANGES AND SHORELINE MIGRATION TRENDS AT THE NILE DELTA PROMONTORIES

O. E. Frihy and A. A. Khafagy
*Water Research Center, Alexandria, Egypt*

## 26.1 Introduction

The Nile is the largest river in the world, its basin covering roughly one-tenth of the area of the African continent (Figure 26.1). Its water comes from two major sources: the Ethiopian tributaries (Atbara and Blue Nile) and the non-Ethiopian, which drains a large area of equatorial east Africa (Sobat River, Bahr El Gazal and Bahr El Gabel). In the delta region north of Cairo, the Nile diverges into two branches that, respectively, form promontories at the Rosetta and Damietta estuaries (Figure 26.2).

Coastal recession has been observed along distinct sectors of the Nile delta coast since the beginning of the present century, and has increased very noticeably during the last three decades. At the same time, accretion has caused shoaling and siltation in the lake inlets and Nile estuaries, leading to navigation problems. Several reports have been published on the Nile delta coastal and nearshore changes, including interpretation of repeated beach profiles (Manohar, 1976), bottom morphology (e.g. Misdorp and Sestini, 1976; Toma and Salama, 1980); and shoreline deviations from topographic maps (e.g. Sestini, 1976; Misdorp, 1977; UNDP/ UNESCO, 1978), satellite photographs (Klemas and Abdel-Kader, 1932; Smith and Abdel-Kader, 1988) and aerial photographs (Frihy, 1988). Until recently, none of the previous studies have considered statistical trend analysis of coastal changes.

### 26.1.1 Effect of dams and barrages on sediment supply

The main Nile and its two branches in Egypt have been controlled by six barrages and three dams (Figure 26.1). The barrages have had only minor effects on the flow pattern during Nile floods, since all gates are kept open. The sluices of the Low Aswan Dam were designed to pass the flood waters and their sediments. However, silting-up of the dam (5 milliard m$^3$ water

301

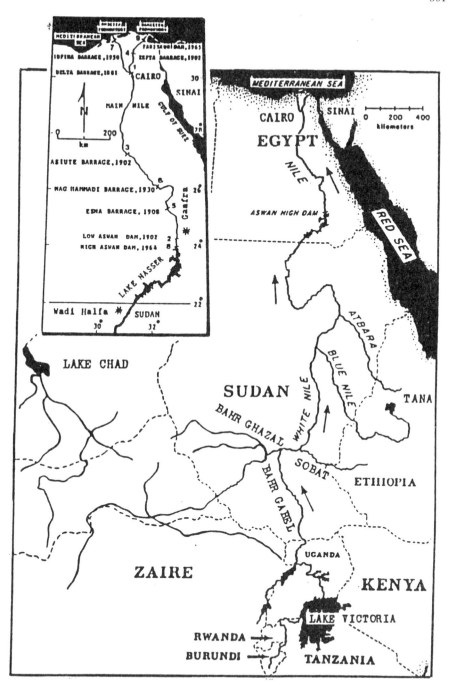

**Figure 26.1** (Left) *Positions and dates of construction for dams and barrages across the Nile in Egypt (numbered 1–9 in chronological order); (right) the Nile's main tributaries and Lake Chad (arrows point downstream)*

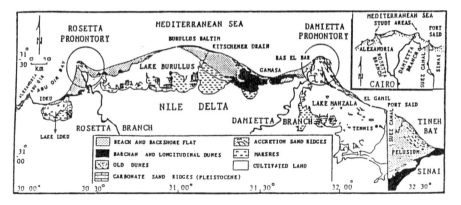

**Figure 26.2**    *Map of the Nile delta coast. Shown are the study areas at the Rosetta and Damietta Nile promontories and the major coastal geomorphological units (after Frihy et al., 1983)*

capacity) had taken place, particularly during flood seasons. The major effect of dams on the shore has been created by construction of the High Aswan Dam, in which all sediments are being trapped and deposited in Lake Nasser (6000 km$^2$), instead of delivering them to the sea through the Nile promontories. The sluices of the High Aswan Dam are opened periodically for the purposes of irrigation, navigation and electric power generation. Sedimentological investigations have indicated that a new delta is being formed in Lake Nasser. Hammad *et al.* (1979) estimated the amount of sand supplied to the beach by the Nile before the construction of the High Aswan Dam to be 19.4 million tons/yr. Most of the literature over the last two decades on the coastal instability of the Nile delta has attributed this phenomenon to the impact of man-made control structures, particularly the High Aswan Dam, which controls the flow of the Nile and thus decreases the annual flood sediment supply to nourish the delta coast. In fact, this common concept coincided with the noticeable coastal erosion and gradual drop in the discharge of the Nile when the Low Aswan Dam was put into operation in 1902. Based on the differences between the annual sediment quantity upstream and downstream of the Low Aswan Dam, a total of 10.5 million tons of sediments were annually trapped in the Low Aswan Dam reservoir (Hammad *et al.*, 1979). Aleem (1972) and Harris (1979) have reported significant differences in the Nile hydrology prior and after the construction of the High Aswan Dam. Harris estimated a decrease in the annual discharge to the sea from 43.5 to 4.4 milliard m$^3$/yr, and a substantial decrease in sediment budget after the construction of the High Aswan Dam.

An important question arises in this situation. Is the decrease of sediment supply to the coast due to man-made water controls or could it be influenced by climatic factors? Such factors could be related to (1) natural changes in the Nile water discharge in historical times, (2) relative sea (land) level changes and (3) the prevailing dynamic forces of winds, waves, currents and tides. This study examines data for Nile flood indicators (water level, sediment load, and discharge) covering the same time span of the shoreline positions studied. The level in the equatorial lakes will be used as a reliable parameter for precipitation. We also report on more recent shoreline positions derived from various sources to study the long-term shoreline migrations during the last two centuries at the Nile promontories. The average rate of erosion will be calculated, based on trend analysis of the outer margins of the Nile promontories. An attempt is made to relate the shoreline migration trends to the available climatic factors, such as historical Nile flood records (sediment load, water level and discharge) in connection with the paleoclimate changes in equatorial East Africa.

## 26.2 Database and analysis

Particular attention has been paid to shoreline changes as part of a long-term plan initiated at the Coastal Research Institute (CRI) in Alexandria to study coastal processes along the Nile Delta coast. The present investigation is based on shoreline surveys conducted by the CRI in 1971, 1981, 1987 and 1988, as well as on two sets of aerial photographs taken in 1955 and 1983. Additional information concerning old shorelines was derived from topographic maps, admiralty charts, and satellite photographs. Sources of shoreline positions used in the present study are listed in Table 26.1. The shorelines of the western and eastern sides of the promontories were matched to the same scale, overlaid and remapped on

**Table 26.1** Data source for historical shoreline positions

| Dates | Data source |
|---|---|
| 1971, 1981, 1987 and 1988 | Ground survey (CRI) |
| 1964, 1935 and 1911 | Ground survey (National Authority of Coastal Protection, 1968) |
| 1955 and 1983 | Aerial photographs (CRI) |
| 1973 and 1978 | Satellite photographs (Klemas and Abdel-Kader, 1982) |
| 1916, 1924, 1942, 1944, 1949 and 1950 | Topographic maps (Egyptian Survey Department) |
| 1800, 1810, 1857, 1864, 1893 and 1909 | Topographic maps (Sestini, 1976; Misdorp, 1977; CRI/UNDP/UNESCO, 1978). |
| 1922 | Admiralty chart |

one chart (Figures 26.3 and 26.4). Long-term rates of coastal erosion are estimated by trend analysis of shoreline positions. The database includes the length (m) measured seaward from a fixed baseline over successive shorelines. In order to calculate the average rate of coastal erosion, several transects were drawn parallel to the shore-migration axis to intercept the shorelines in the study areas. Across these lines, the distance from interceptions to the baseline were measured and averaged for each shoreline position. The numerical data sets of length for each shoreline were subjected to least-squares regression analysis in order to calculate the rate of erosion during periods of regression.

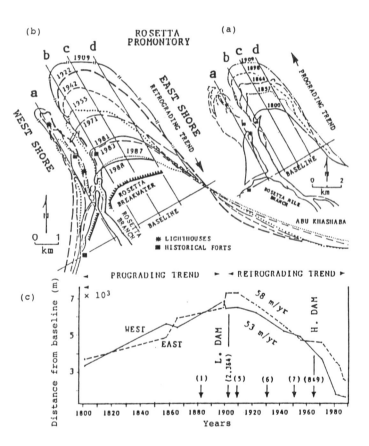

**Figure 26.3**   *Shoreline changes at the western and eastern Rosetta promontory, illustrating the selected transects (a–d) used for measuring the change in shoreline positions from the baseline during the nineteenth and twentieth centuries. Arrows on the horizontal axis indicate dams and barrages shown in Figure 26.1. (a) The prograding shorelines; (b) the retrograding shorelines; (c) trend of shoreline positions in metres as measured from the baseline for the western and eastern sides of the promontory*

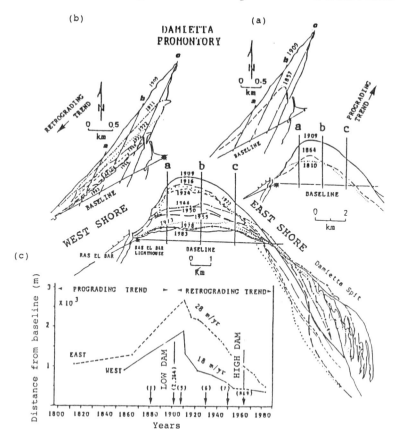

**Figure 26.4** *Shoreline changes at the western and eastern Damietta promontory (for details, see Figure 26.3)*

Historical records of Nile discharge from 1871 to 1971 are extracted from documents published by CRI/UNDP/UNESCO (1978). More recent data from 1972 to 1988 were obtained from the Ministry of Public Works and Water Resources. The data set was subjected to time-series analysis and smoothed by four-mean averaging.

## 26.3 Discussion

The overall shorelines at the Rosetta and Damietta promontories exhibit two phases of long-term shoreline migrations. The data indicate that the shoreline built up during the nineteenth century and receded over the twentieth.

### 26.3.1 Rosetta promontory

The successive shorelines west and east of Rosetta allow for the analysis of their positions back to 1800 (Figure 26.3(a) and (b)). The shoreline has changed markedly in position and configuration during the 188 years. Between 1800 and 1909 the promontory advanced about 3.5 km, whereas the shoreline pattern from 1909 to 1988 on the tip of the promontory shows a considerable retreat on both the western and eastern sides of the Rosetta mouth. On the other hand, an accretionary shoreline is detectable on its eastern side at the Abu Khashaba embayment (Figure 26.3(b)). This is attributed to sand eroded from the outer margin of the promontory being transported by waves and longshore current to the cast. Longshore currents at the Rosetta promontory flow eastward with a 55% frequency have been reported by Fanos (1986).

It would appear that the long-term rates of erosion estimated from regression analysis of the length changes in the western area (53 m/yr) are comparable to the eastern part (58 m/yr). The rate of erosion at the eastern part represents the highest shoreline regression documented along the Nile delta coastline. The retrograding trend also shows a significant rapid shoreline retreat from 1970 until 1988.

### 26.3.2 Damietta promontory

The shoreline trend at the Damietta promontory over the nineteenth and twentieth centuries is shown in Figure 26.4. The shoreline changes are similar to those at the Rosetta promontory in which a prograding trend occurred in the nineteenth century while a retrograding one existed during the twentieth. The magnitude of the shoreline retreat at the western side is small (18 m/yr) compared with that of the eastern side (28 m/yr). There is an accretionary pattern east of the Damietta mouth as recorded by the growth of a complex spit. The spit was observed on the 1973, 1978 and 1983 shorelines, and shows considerable longshore development from 1971 to 1983. The shoreline pattern of the Damietta promontory indicates that at least part of the sediment eroded from its outer margin has been transported towards the east in the form of the Damietta spit. Inman *et al.* (1976) estimated the wave-induced longshore sediment transport rate at the Damietta promontory to be eastward at about $860 \times 1000$ m³/yr.

### 26.3.3 Shoreline trends and climatic fluctuation

As for other of the world's deltas, the major factors controlling the coastal changes of the Nile delta are sediment supply, rise in sea level and dynamic forces. Here we explore the connection between the migration of the shoreline, the influence of the Nile flood stages during the same time span, and climatic changes in Africa. Several independent studies have reported on long-term series of Nile flood parameters (sediment load, water level and discharge). Work was carried out by CRI/UNDP/UNESCO (1978) on the fluctuation of the Nile sediment load from historical Nile gauge records. Five-year moving averages related to the estimated suspended sediment load at Gaafra, situated 35 km downstream from the Low Aswan Dam are shown in Figure 26.5(a), covering the period

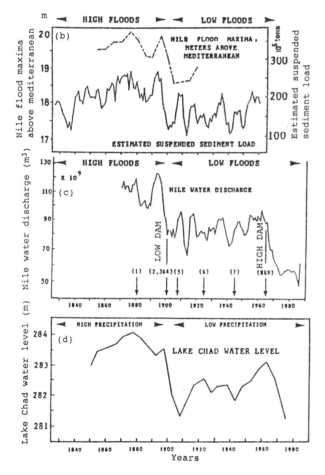

**Figure 26.5** *Trend of Nile flood stages and Lake Chad water level over the previous two centuries. Arrows denote the construction dates of dams and barrages along the Nile (numbered 1–9 and shown in Figure 26.1). Note the chronological similarity in trends among curves (a)–(d) as well as the discontinuity around the year 1900. Curve (a) – distribution of sediment suspended load at Aswan from 1825 to 1963 (after CRI/UNDP/UNESCO, 1978); curve (b) – distribution of the Nile's water levels above the Mediterranean at the Roda nilometer from 1850 to 1920 (after Hassan, 1981); curve (c) – smoothed annual distribution of the Nile's water discharge at Aswan from 1871 to 1988 (data obtained from CRI/UNDP/ UNESCO, 1978, and the Ministry of Public Works and Water Resources); curve (d) – Lake Chad water level (Maley, 1973; cf. Hassan, 1981)*

1825–1963. A significant difference can be observed in the suspended load prior to and after 1900. It appears that the Nile suspended load was higher (on average, by 200 million tons/yr) in the nineteenth century (indicating high floods) than in the twentieth (on average, 160 million tons/yr), indicating low floods. A similar trend has been reported by Hassan (1981), using historical records of Nile flood stages in terms of Nile water levels above the Mediterranean, from 1850 to 1920 (Figure 26.5(b)). These data are based on Nile water levels measured by others at the Roda Nile gauge (nilometer) near Cairo, which partially covers the last two centuries. It also reflects the behaviour of the Nile levels, which were particularly high in the nineteenth century and significantly lower in the twentieth. Hassan (1981) related the fluctuation of Nile floods to the climatic changes in East Africa. The same conclusion has been also reached by Riehl and Meitin (1979), who found a significant climatic variation of the order of 100 years in the historical annual Nile discharge dating from the years 622 to 1976. A series of years of high discharge (1870–1900) was followed by a series with low discharge, almost suggesting a discontinuity. On average, the discharge is 109 milliard $m^3$ for the period 1800–1899 compared with 83 milliard $m^3$ for 1900–1986.

Here we also present a time series for Nile discharge records at Gaafra (35 km downstream from Aswan) covering the period from 1871 to 1988 (Figure 26.5(c)). On average, the annual discharge is estimated to be 108 milliard $m^3$/yr in the period 1871–1902 and 79 milliard $m^3$/yr during 1903–1988. Again the two segments of high and low floods prior to and after 1900 can be seen. The pattern of water discharge seems to correlate well with the other previously mentioned curves (Figure 26.5(a) and (b)).

Earlier studies by CRI/UNDP/UNESCO (1978) have indicated a considerable episodic variation in the magnitude of the Nile water discharge in the nineteenth and twentieth centuries at Wadi Halfa (upstream), on the Egyptian-Sudanese border and as well as at Aswan (downstream). These differences were attributed to climatic factors. Since the change in water discharge and sediment loads were based on data collected downstream from the Low Aswan Dam, and on the conclusion arrived at by CRI/UNDP/UNESCO (1978), the Nile sediment discharge may have been affected by the Low Aswan Dam as well as climatic changes.

The water levels in some equatorial lakes are mainly a function of rainfall, and so can also be used as a parameter to indicate climatic changes. Jakel (1984) reported a similar trend between climatic fluctuations in the Sahelian zone in Africa and the water levels of Lake Chad (25 000 $km^2$). The changing water levels of Lake Chad are shown in Figure 26.5(d) (Maley, 1973, cf. Hassan, 1981). It is interesting to note the close consistency between that level and episodic variations in the Nile flood discharge and consequently in sediment load. This indicates that years of high Nile floods have been associated with intense tropical monsoon rainfall. It may thus be inferred that the episodic variations in Nile floods are probably the result of climatic changes that influence precipitation in Africa, i.e., the changes in discharge and sediment loads in the first half of the twentieth century were caused by climate changes rather than dams. The causes of climatic changes in Africa are far from clear, and have been discussed by Tanaka *et al.* (1975) and Fairbridge (1984). The correspondence between the low Nile floods in the twentieth century and the low water levels in Lakes Chad and Victoria may confirm that the retrograding shoreline trend at the Nile delta promontories has been controlled by long-term climatic causes in addition to the effects of the artificial water control system along the Nile. To confirm this relationship, values of Nile water discharge and shoreline changes for corresponding decades are subjected to correlation coefficient analysis (Figure 26.6). The results show correlation coefficient values of +0.7 and +0.5, respectively, for the west and east sides of the Rosetta promontory, while at the west and east

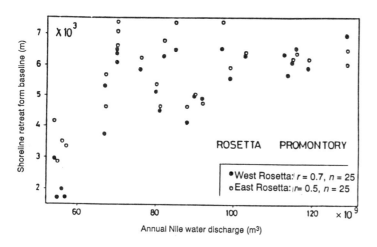

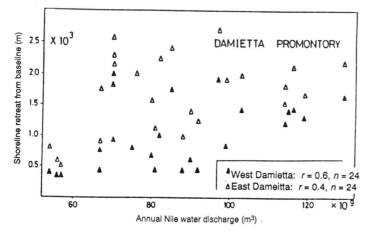

**Figure 26.6** *Comparison of annual Nile water discharge and shoreline retreat at the Nile promontories. Results of correlation coefficient indicate significant correlation*

Damietta promontory they are +0.6 and +0.4, respectively. These correlations are statistically significant at the 95% confidence level. This analysis therefore strongly confirms the relationship between Nile floods and shoreline changes. This proves that there is a close connection between historical changes in the Nile flood stages and the shoreline trends studied here.

## 26.4  Conclusions

This investigation, based largely on the positions of recent and historical shorelines, defines two major shoreline migration trends that have dominated the Nile delta promontories during the previous two centuries. A prograding trend occurred during the nineteenth century whereas a retrograding one developed in the twentieth. The results of shoreline retreat estimated by trend analysis show a high rate of erosion along the tip of the Rosetta promontory (53–58 m/yr), while a relatively low rate occurs at the tip of the Damietta promontory (18–28 m/yr). Significant accretionary patterns are also linked with shoreline retreat, as has been documented on the eastern side (saddle) of the Rosetta mouth and to the east of the Damietta promontory (spit zone formation).

The study also reveals a close relationship between the determined shoreline migration trend and the historical changes in the Nile water flood stages as a result of the effect of global climatic changes, which could be of an overall cyclic nature, and are probably due to fluctuations in monsoonal rainfall in east Africa. This may illustrate the concept that the prograding shoreline pattern was built up during an episode of relatively high floods (greater rainfall) in the nineteenth century, while periods of low Nile discharge were probably the prime factor responsible for delta shores retrograding prior to the total damming of the Nile by the High Aswan Dam. In general, climatic changes as well as man-made water control measures have been the main factors in Nile delta erosion.

## Acknowledgements

We are indebted to Dr El Askary, Alexandria University, and Dr P. Komar, University of Oregon, for their constructive review of the manuscript. Useful comments and suggestions were also provided by Dr G. Sestini, UNEP.

# References

Aleem, A. A. (1972) 'Effect of river outflow management on marine life', *Mar. Biol. J.*, **15**, 200–208.

CRI/UNDP/UNESCO (1978) *Coastal protection studies*, Final Technical Report (Paris), **1**, 155.

Fairbridge, R. W. (1984) 'The Nile floods as a global climatic soil proxy', in Morner, A. and Karfen, W. (eds), *Climatic Changes on a Yearly to Millennial Basis*, Reidel, Dordrecht, pp. 181–190.

Fanos, A. M. (1986) 'Statistical analysis of longshore current data along the Nile Delta coast', *Water Science Journal, Cairo*, **1**, 45–55.

Frihy, O. E. (1988) 'Nile Delta shoreline changes: Aerial photographic study of a 28-year period', *J. Coastal Research, USAV*, **3**, 597–606.

Frihy, O. E., El Fishawi, N. M. and El Askary, M. A. (1988) 'Geomorphological features of the Nile delta coastal Plain: A review', *Acta Adriat., Yugoslavia*, **29**, 51–65.

Hammad, H. Y., Khafagy, A. A., Mobarak, I. and Sidky S. (1979) 'A short note on the sediment regime of the Nile river', *Bull. Inst. Ocean. & Fish., ARE*, **7**, 314–322.

Harris, F. R. (1979) *Master planning and infrastructure development for the Port of Damietta*, Consultant Report to Ministry of Reconstruction and New communities, Egypt.

Hassan, F. A. (1981) 'Historical Nile floods and their implications for climatic changes', *Science*, **212**, 1142–1145.

Inman, D. L., Aubrey, D. G. and Pawke, S. S. (1976) 'Application of nearshore processes to the Nile Delta, a preliminary report', *Proceedings UNESCO Seminar on Nile Delta Sedimentology*, Alexandria, Egypt, pp. 205–255.

Jakel, D. (1984) 'Rainfall patterns and lake variations at lake Chad', in Morner, N. A. and Karlen, W. (eds), *Climatic Changes on a Yearly to Millennial Basis*, Reidel, Dordrecht, pp. 191–200.

Klemas, V. and Abdel-Kader, A. M. (1982) 'Remote sensing of coastal processes with emphasis on the Nile Delta', in *International Symposium on Remote Sensing of Environments*, Cairo.

Manohar, M. (1976) 'Reach profiles', *Proceedings UNESCO Seminar on Nile Delta Sedimentology*, Alexandria, pp. 95–99.

Misdorp, R. (1977) 'The Nile promontories and the Nile continental shelf', *Proceedings UNESCO Seminar on Nile Delta Coastal Processes*, Alexandria, pp. 456–551.

Misdorp, R. and Sestini, C. (1976) 'Notes on sediments map of the *Endeavour* survey of 1919–1922', *Proceedings UNESCO Seminar on Nile Delta Sedimentology*, Alexandria, pp. 191–204.

National Authority of Coastal Protection (1968) *Maps showing the coastal changes along the Northern coast*, Cairo (in Arabic).

Riehl, H. and Meitin, J. (1979) 'Discharge of the Nile River: a barometer of short-period climate variations', *Science*, **206**, 1178–1179.

Sestini, G. (1976) 'Geomorphology of the Nile Delta', *Proceedings UNESCO Seminar on Nile Delta Sedimentology*, Alexandria, pp. 12–24.

Smith, E. S. and Abdel-Kader, A. (1988) 'Coastal erosion along the Egyptian Delta', *J. Coastal Research*, **2**, 245–255.

Tanaka, M., Weare, B., Navato, A. and Newell, B. (1975) 'Recent African rainfall patterns', *Nature*, **255**, 201–203.

Toma, S. A. and Salama, M. S. (1980) 'Changes in bottom topography of the western shelf of the Nile Delta since 1922', *Mar. Geol.*, **36**, 325–339.

# 27 CLIMATIC INFLUENCE ON RECHARGE AREAS, DISCHARGE AND AQUIFER DIFFERENTIATION IN LAHNSTEIN-ON-RHINE, FRG

M. Fricke
*Hydrogeo Consulting Ltd, Bad Driburg, Germany*

In 1986 a large German brewery bought a mineral water-bottling plant in Lahnstein-on-Rhine, near Koblenz, and geologically in the Middle-Rhine and Graben Valley system. Victoria Heil- und Mineralbrunnen GmbH is an old and well-known mineral water company, founded in 1879, with special technical equipment and a knowledge of springs and wells, originally exporting mineral water to the Dutch courts.

When dealing with any kind of water in Germany, the first thing to do is to check the water rights: quantity, quality and expiry. You are not allowed to produce or use any water in Germany without a permit, which is normally valid for 25–30 years, and is granted by the government. According to German water legislation, you are not allowed to extract more water from a surface or subsurface water resource than is naturally recharged. This ensures that there is no over-exploitation of water deposits or groundwater extraction in Germany. In a 'Claim for a Water Right' a hydrogeologist must prove that there is no potential over-exploitation. Consequently a water right for, say, 25 years, is always limited to a stipulated quantity. When owners of a mineral water bottling plant intend to increase discharge from their wells they must submit a 'Claim for a New Water Right'.

## 27.1 Recharge – discharge

As the discharge of a well or spring is always directly or indirectly linked to the recharge (except under certain hydrogeological conditions – e.g.

natural waters) it is the task of every consulting hydrogeologist to discover if there have been changes in the last 20 or 30 years in the recharge area of the wells or springs (Figure 27.1). Therefore the author began to collect weather data for the potential recharge area of the Victoria Heil- und Mineralbrunnen GmbH.

In Germany it is easy to obtain rainfall data for any region due to an efficient nationwide weather service. The German Weather Service in Essen provided the following data for the recharge area of the Lahnstein wells (Figure 27.2):

Annual rainfall, Bad Ems: 697 mm (north-east Lahnstein)
Annual rainfall, Lahnstein: 585 mm
Annual rainfall recharge area: 660 mm

It is far more difficult to obtain reliable evaporation and surface discharge data in order to calculate the underground discharge, i.e. the amount of underground water recharging the wells. The underground discharge is one of the most important parameters in regard to the wells' discharge in relation to the underground recharge. Following German water legislation, the wells' discharge must not be more than the underground recharge in order to avoid over-exploitation or irreversible damage to the water tables.

An evaporation of about 45% may be estimated and a surface discharge of about 15% of average annual rainfall, i.e..

$$U = R - E - S$$

where

$U$ = underground discharge
$R$ = rainfall (660 $l/m^2/yr$),
$E$ = evaporation (297 $l/m^2/yr$),
$S$ = surface discharge (99 $l/m^2/yr$).

The underground discharge/recharge of the wells is 264 $l/m^2/yr$, which, applied to 21 $km^2$ of surface recharge area, equals 5 544 000 $m^3/yr$. Discharges of the wells were as follows:

Louise      : 411 720 $m^3/yr$ = 7.42%
Adele       : 438 000 $m^3/yr$ = 7.90%
Minerva 1 :  21 000 $m^3/yr$ = 0.35%
Minerva 2 : 131 400 $m^3/yr$ = 2.37%

Only 18.04% of the underground recharge is affected by the total discharge of all the four wells. Therefore the water authorities may agree that there is no danger to the water balance in this area.

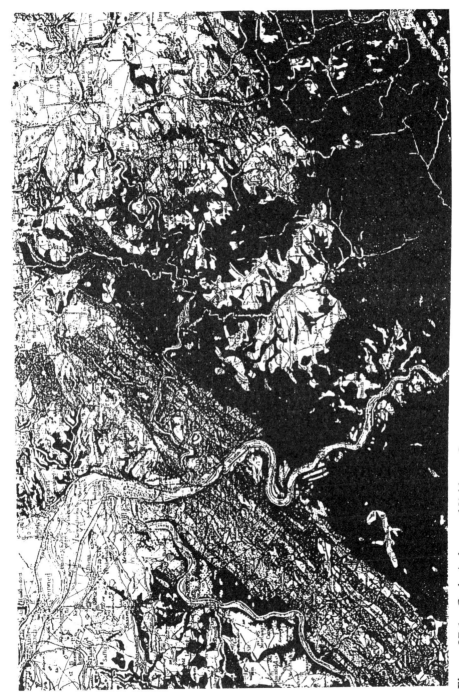

**Figure 27.1** *Geological map of Koblenz, Germany*

Medium annual/monthly rainfall

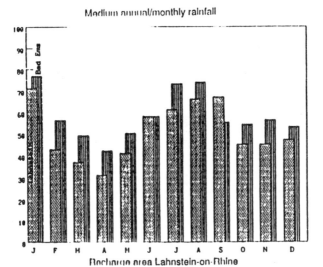

Recharge area Lahnstein-on-Rhine

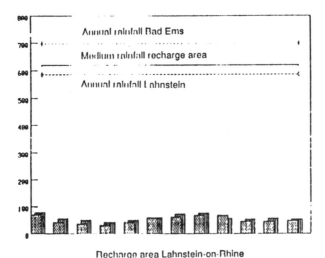

Recharge area Lahnstein-on-Rhine

**Figure 27.2**

## 27.2 Two aquifers?

New hydrogeological and geophysical research raised a problem: Might there be two aquifers and therefore two different recharge areas? Until recently, all the geologists who had worked in the interesting Emser-Spring-Sattel area since 1915 were certain that there was only one Lower Devonian aquifer. However, the new results of a geological and geophysical survey revealed a slightly different situation – at least locally.

In the summer of 1988, all the wells of the Victoria Heil- and Mineralbrunnen GmbH based in Lahnstein were subject to stocktaking. In this process, a geophysical survey was conducted on Minerva 1. Up to that time, very little had been known about this well. There were fears, however, that damage in the structure of this well might have adverse effects on the Minerva 2 well located 15 m south of Minerva 1, and perhaps also on the Louise well, an officially approved medicinal spring.

Minerva 1 was probably sunk in 1911 by a Dutch drilling company, and was believed to have a depth of about 50 m. There were no drawings of the well's structure or design, no geological descriptions of the well section, no hydrogeological evaluations of pumping tests, and no chemical analyses.

## 27.3 Geological background

The well had been drilled in the south-eastern flank of the so-called Oberlahnsteiner Sattel, a saddle extending from Neuheusel (Westerwald) to Oberlahnstein. A very large trough-shaped valley connects the Oberlahnsteiner Sattel with the Emser Quellsattel, a spring saddle to the north-east (Figure 27.3). The oldest strata in this region were found in the Adele well at a depth of approximately 420 m: a bed of Lower Devonian Emsian quartzite. Dr Heyl from the Geologisches Landesamt of the Rhineland Palatinate provided the author with the only reliable description of core samples available (Mestwerth, 1934/1936).

The next strata up, with a thickness of about 200–250 m, are the so-called Horrenrheiner Schichten, a sequence of laminated quartzite and quartzitic graywacke slate layers with an embedded rock bind. This is followed by the Laubach Schichten, a ribboned slate bed with interjacent plate-shaped graywacke formations. The top layer of this rock sequence in the region of the medicinal and mineral springs of Lahnstein is formed by Rhine lower terrace gravel and sand.

The Lower Devonian strata in this region are subject to a considerable tectonic load, which has had widely varying effects, depending on the rocks' petrographic character. More clayey complexes have become highly slaty, while hard and compact rocks show clear evidence of cleavage which, in the final analysis, is responsible for the delivery of water and the migration of carbonic acid.

Against the background of this geological formation the 'unknown' well was expected to show over its 50 m of depth ribboned slate, laminated graywacke and – by analogy with Mestwerth's description of the Adele well – perhaps also some hard sandstone. If the support of this well was found to be defective and corroded, this would not represent an immediate risk for the Louise medicinal spring; however, it would jeopardize the Minerva 2 well, located close by.

The first survey of the well (by plumbing its depth) produced a rather surprising finding: the depth was not 50 m, nor was it 190 m, as suspected

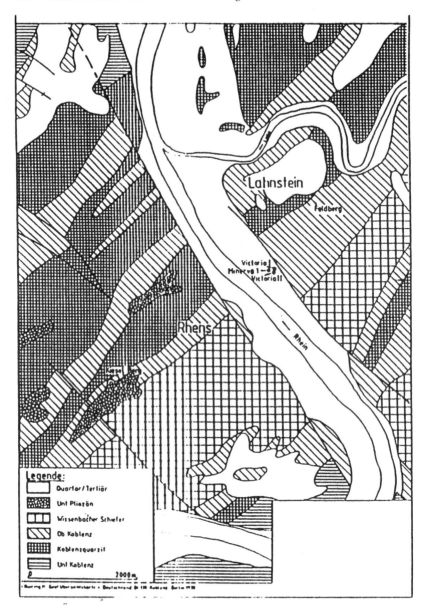

**Figure 27.3**   *Geological map of Lahnstein and its environs*

by some of the company's long-standing employees; it was much deeper than that. Plumbing showed that the depth of the well was 220 m below the surface. With this information, it was imperative to carry out an accurate geophysical survey on the well because there was a possibility that, at 220 m, it had been drilled into the upper strata of the Hohenrheiner Schichten, i.e. the upper parts of the aquifer which was very likely to be the source of water feeding the Louise medicinal spring. This meant that there was a possibility that a defective well structure would affect the officially approved medicinal spring.

## 27.4 Geophysical survey (Figure 27.1)

The survey involved measurement of the following parameters:

1. Natural gamma radiation (*GR*)
2. Electrical resistance of strata (*FEL*)
3. Hole calibre
4. Relative vertical flow rate (*FLOW*)
5. Well water temperature (*TEMP*)
6. Electrical conductivity of well water (*SAL*)

The first three parameters are usually measured in open drill holes, providing information on strata properties. However, they may also be used in finished wells to obtain data on the well's support and condition. Parameters (4)–(6) are typically measured in finished wells, providing information on water inflow and outflow. This chapter will concentrate on the latter measurements, whose results are illustrated in Figure 27.4.

Well water measurements are usually carried out while the probe is being lowered into the well and not while it is hoisted back up. This prevents the probe and the measuring cable from affecting the water to be tested. At the time when the measurements were made, water was pumped up from a depth of about 18 m. As a general rule, it is not possible for the measuring probe to pass the pump without being subject to interference. The data obtained by this measurement are only valid, therefore, for the area below the pump (i.e. at −20 m). The temperature plot (*TEMP*) on the right-hand side of Figure 27.4 shows that the temperature rises slowly with increasing depth. There is no indication of any inflows or outflows of water.

Electrical conductivity (*SAL*) shows a gradual rise up to a depth of approximately 203 m and then begins to increase very rapidly. The abnormal readings between −120 and −126 m, and between −143 and −146 m, are due to the fact that the probe became stuck while it was lowered into the well. The sudden rise in conductivity below a depth of 120 m is most probably due to a change in the well's support. Below a depth of 203 m, electric resistance readings and gamma radiation intensity also show a decrease and an increase, respectively (i.e. the support ends at 203 m below surface level). This finding is corroborated by the flow velocity data measured (*FLOW*). They are shown by the first two diagrams of Figure 27.4.

Water flow velocity is measured by means of a radial vane meter which records the vertical relative motion between the well fluid and the probe.

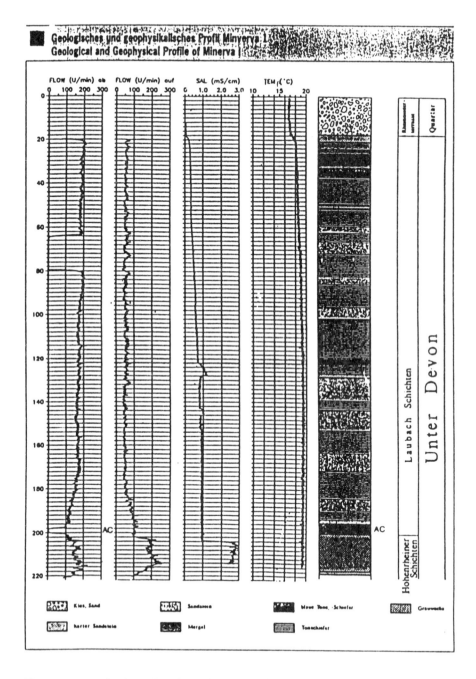

**Figure 27.4**   *Geological and geophysical profile of Minerva 1*

The speed of the radial vane, which is roughly proportional to the amount of relative velocity, is measured in revolutions per minute (rpm):

$$f = k \, | \, V_S - V_F \, | \text{ (rpm)}$$

where

$f$ is the radial vane speed (rpm),
$k$ is the device constant (rpm),
$V_S$ is the probe travel speed (m/min),
$V_F$ is the well fluid flow rate.

The velocity readings on the meter are therefore independent of the direction of rotation of the radial vane, since this only records the amount of relative velocity.

The first diagram in Figure 27.4 shows the data measured when the probe was lowered into the well. It is for this downward movement that the meter had been designed. Hence, it responds very sensitively during an upward movement. At a depth of between 64 and 80 m, the vane was blocked by suspended solids in the water, therefore there are no velocity readings for that depth range. The same phenomenon may have occurred at a depth of between 198 and 200 m. Another explanation which cannot be ruled out, however, is that it is because of inflow or outflow of water that the relative flow velocity between the probe and the water drops below the minimum at that level so that there is not enough movement to drive the vane. The gradual reduction of this effect between approximately 170 and 198 m suggests that there is a small amount of inflowing or outflowing water.

On the way up, the probe showed slightly different results. Meter deflections as plotted in the diagram were highest and most pronounced at a depth of greater than 203 m. Velocity readings begin by rising rapidly, eventually showing a gradual decrease up to 174 m. At a depth of 20 m, the curve is then almost at a constant level.

The effect recorded by this metering instrument is a combination of two velocities: the probe's travel speed plus the vertical flow rate of the water. On the way up, the vane speed recorded at the end of the well while the probe was stationary was approximately 100 rpm (i.e. at this level there is a constant water flow rate). When the probe was accelerated to its normal travel speed of 6 m/min, the meter velocity reading was above 200 rpm (i.e. at this level the direction of probe travel is opposite to that of water flow, which is downward). It is only at a depth of 203 m that the amplitude of probe deflections decreases and the vane speed is what it was at the bottom of the well while the probe was stationary (i.e. 100 rpm). Therefore the difference between the probe's speed of travel and the water flow rate is equivalent to the flow rate of the water at the bottom of the well.

This also proves that the vane was not blocked by suspended well water solids at a depth of between 198 and 200 m, when the probe was lowered into the well, but that the relative velocity dropped below the minimum required to keep the vane in motion. This suggests that the descending probe and the ambient water move with almost the same velocity (6 m/min), although water is being pumped upwards. Below a depth of 200 m, the velocity of the downward water flow is greater than that of the descending probe, so that in fact, there should be negative meter readings. These, however, are not admissible according to the formula for velocity of rotation (because of the amount of differential velocity).

## 27.5 Conclusions

### 27.5.1 Geophysical findings

Geophysical measurements in Minerva 1 – a previously 'unknown' well – provided evidence of a rather unusual phenomenon, i.e. that at lower depths, there is a downward water flow, while at the same time water is pumped to the surface from a level of about 18 m, thereby generating an upward flow. The velocity of the downward water flow is not negligible, since it exceeds the probe's normal descent speed of 6 m/min.

### 27.5.2 Geological/hydrogeological findings (Figure 27.5)

The geophysical findings obtained at a depth of between 198 and approximately 200–203 m coincide with (and confirm) Mestwerth's observations. In his descriptions of core samples, Mestwerth noted a sequence of dense, compact clay layers – so-called 'blue clay' – at a depth of between 198 and approximately 202 m. If one relates the geophysical findings to the observations made in 1934/1936, this suggests that the blue clay must be an aquiclude, which at this point in the saddle slope of the Oberlahnsteiner Sattel separates two groundwater storeys. The high water flow rates in the lower part of the well indicate that there is a substantial pressure differential between the two Lower Devonian aquifers, and that the lower one is subject to considerable 'drainage'. Without doubt, these findings prove that the Minerva 1 well does, indeed, represent a major risk to the officially approved Louise medicinal spring. It can almost safely be assumed that the Minerva 1 well was drilled into the upper part of the aquifer which feeds the Louise spring, and that it perforated the aquiclude protecting these water resources. Furthermore, the survey showed that between a depth of 174 and 198 m the well casing is defective and corroded.

It is not possible in this context to determine whether the sealing effect of the 'blue clay' has had any impact on carbonic acid migration, since although the highly complex Rhine Graben tectonics, in terms of jointing

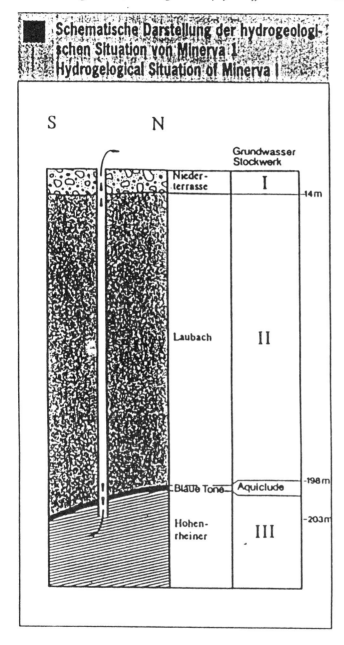

**Figure 27.5** *The hydrogeological situation of Minerva 1*

and currents, have a very positive effect on carbonic acid migration, it is very difficult to detect the exact pathways of this migration.

When the management of the wells' operating company was made aware of these geophysical results, comparison with the geological/ hydrogeological observations and evaluation of the findings, they decided to seal the Minerva well in order to protect their medicinal spring and to preserve their underground water resources.

### 27.5.3 Recharge area

Comparisons with two other boreholes in the area of Bad Ems showed that those 'Blue Schists' do not exist in the main recharge area. Therefore it is unlikely that there is a separation of two Lower Devonian aquifers in the recharge area and the aquifer differentiation is only of local importance, the upper aquifer producing a low mineralized water, the deeper one a highly mineralized water. The recharge area, consisting of a fissured Lower Devonian aquifer, represents a single aquifer and no division into two recharge areas must be made.

# References

Carle, W. (1975) 'Die Mineral- und Thermalwässer von Mitteleuropa', *Wiss. Verlagsgesellschaft*, Stuttgart.

Dillmann, W. (1974) 'Abgrenzung eines Heilquellenschutzgebietes zu Gunsten der Victoria Brunnen AG, Lahnstein', *Geol. Landesamt Rheinland-Pfalz*.

Fricke, K. and Michel, G. (1974) 'Mineral- und Thermalwässer der Bundes-republik Deutschland', *Mineralbrunnen*, **23**, 3–14, Bad Godesberg.

Fricke, M. (1987) *Die Erschließung von natürlichem Mineralwasser in industriellen Ballungsgebieten*, Report I, SITH Cong., Karlovy Vary.

Heyl, K. E. (1972) *Bäderbuch von Rheinland-Pfalz*.

Mestwerth, W. (1934–1936) 'Bohrprobenbeschreibung eines neuen Brunnens in Lahnstein', *Bohrarchiv Geol.* Landesamt Rheinland-Pfalz.

Quiring, H. (1930) 'Geolog. Übersichtskarte von Deutschland', *Blatt Koblenz* No. 138, Preussische Geologische Landesanstalt, Berlin.

# 28 EL NIÑO AND ITS EFFECT ON THE INDIAN CLIMATE

S. N. Kathuria and S. D. Kulkarni
*Central Water Commission, New Delhi, India*

## 28.1 Introduction

The world's climate has always been changing, and it is reasonable to assume that it will continue to do so in the future (Lamb, 1977). The climate of the past million years is revealed by the ecological interpretation of paleobiological materials (remains of plants). The overall climate is changing compared to that of millions of years ago. It is generally believed that it is becoming cooler compared with 20 million BP. The present century showed a warming trend during the 1920s and 1930s followed by cooling up to the 1960s, and then reversing after the late 1960s (Wallen, 1984).

It is generally believed that El Niño is one of the major factors responsible for global climate fluctuations. In this chapter an attempt is made to study the theoretical aspects of El Niño and its effect on India's south-west monsoon.

In 1924, Sir Gilbert Walker observed that the vagaries of the monsoon rainfall are highly correlated to the phenomenon known as the Southern Oscillation (SO). It has been regarded as a large-scale seesaw of atmospheric mass between the Pacific and Indian Oceans in the tropics and sub-tropics. The seesaw of atmospheric pressure involves development of low- and high-pressure zones in the Indian and Pacific Oceans, respectively, and vice versa. The oscillation is irregular, and its period of occurrence lies in the range of 2 to 7 years. Various kinds of indices are used to study the influence of this phenomenon.

El Niño is another event which also belongs to the Southern Oscillation family and which has a significant influence on Asia's monsoon circulation. This is generally referred to as an anomalous warming of the equatorial eastern Pacific waters off the Peru coast. It has the global effect of flooding in Ecuador, Cuba, etc. and causing droughts in Asian countries. Weak warm currents off the coast of Peru and Ecuador are a general annual phenomenon, and this is termed the current of Christ (or child), as it develops around Christmas every year. It is only when this becomes intensified that it is known as 'El Niño'.

## 28.2 Occurrence of past events

From ship observations over the period 1875–1980 a total of 25 mod-
erate/strong events were identified (Quinn *et al.*, 1978, Rasmusson and
Carpenter, 1982) – an average of one every 4.2 years. The identification of
moderate and strong warm episodes was subjective, but, in retrospect,
they include all events which exhibited a maximum positive anomaly
$\geq +1.0°C$ and a change from the largest negative anomaly of the previous
year to the largest positive one of the warm-episode year (+2.5°C). Two
more El Niños were identified (1982–1983 and 1986–1987). A list of
warm-episode years is given in Table 28.1.

## 28.3 Meteorological features

The general features of El Niño are as follows (Ramage and Hori, 1981):

1. El Niño is associated with the South Pacific trade winds, which are
   relatively weaker than those of the North Pacific, especially in tropical
   east Pacific coastal waters.
2. See surface temperatures (SSTs) are generally well above normal along
   the equator and off South America, and positive anomalies may extend
   beyond 10°N and 10°S.
3. El Niño generally begins in March or April and may last for a year or
   more. When positive sea surface temperature anomalies decrease between June and September, El Niño weakens.
4. El Niño events are associated generally with strong surface winds
   except the band between 7°N and the equator.
5. The ocean gains maximum heat where the SST is lowest and much less
   when the SST is highest.
6. El Niño is associated with maximum surface wind convergence and
   highly reflective cloud frequency at the latitude of zero meridional
   surface wind (Kilonsky and Ramage, 1976).
7. Rainfall is concentrated in the zone of maximum surface wind convergence and can be accounted for in terms of horizontal convergence and
   upper air divergence (stronger in El Niño).
8. There is no clear relation between El Niño and tropical cyclone development.
9. In tropical regions an anomalous warm ocean may be expected to
   produce extra convection in the air above (Hoskins and Simmons, 1979).

**Table 28.1**   Warm-episode (El Niño) years

| S no. | Year | Source |
|-------|------|--------|
| 1 | 1877 | Q |
| 2 | 1880 | Q |
| 3 | 1884 | Q |
| 4 | 1887 | Q |
| 5 | 1891 | Q |
| 6 | 1896 | Q |
| 7 | 1899 | Q |
| 8 | 1902 | Q |
| 9 | 1905 | Q |
| 10 | 1911 | Q |
| 11 | 1914 | Q |
| 12 | 1918 | Q |
| 13 | 1923 | Q |
| 14 | 1925 | R |
| 15 | 1930 | R |
| 16 | 1932 | R |
| 17 | 1939 | Q |
| 18 | 1941 | Q |
| 19 | 1951 | R |
| 20 | 1953 | R |
| 21 | 1957 | R |
| 22 | 1965 | R |
| 23 | 1969 | R |
| 24 | 1972 | R |
| 25 | 1976 | R |
| 26 | 1982 | P |
| 27 | 1986 | IMD |
| 28 | 1987 | IMD |

Q = Quinn *et al*. (1978).
R = Rasmusson and Carpenter (1983).
P = Philander (1981).
IMD = India Meteorological Department Report
(1988).

## 28.4  Development of El Niño

The development of a typical El Niño event can be divided into three
phases (Philander, 1983):

1. A precursor phase before the onset of El Niño off the South American
   coast in the spring;

2. A development phase during which anomalous conditions increase; and
3. A decaying phase in which El Niño decays and normal conditions prevail.

### 28.4.1 Precursor phase

According to the Walker Circulation, the prevailing westward trade winds over the tropical Pacific converge on the low-pressure zone of Australia–Indonesia. The air here rises vertically causing much cloud and rainfall. It returns to higher altitudes and then sinks, forming the cold, dry south-east Pacific High-Pressure Zone. The onset of El Niño is preceded by the eastward displacement of the upward part of the Walker Circulation, as revealed by records of surface pressure, winds and rainfall. The onset of El Niño may be initiated by the increase of surface pressure at Darwin, Australia, with a decrease in trade winds and rainfall over Indonesia (Figure 28.1).

In addition to a shift in the Walker Circulation, the Inter-Tropical Convergence Zone (ITCZ), the narrow zonal band of rising air, cloud and high rainfall where south-east and north-east trade winds meet shifts seasonally between 10°N in August and September to 3°N in February and March. The ITCZ is displaced southwards in the east Pacific during the early months of El Niño. This displacement is associated with weak winds

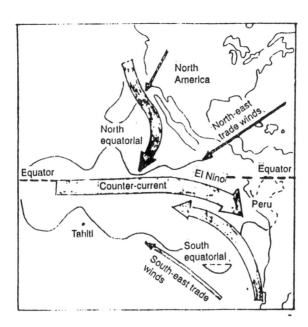

**Figure 28.1**   *The generation of El Niño*

and unusually high sea-surface temperature in the south-eastern equatorial Pacific (Wyrtki, 1975).

### 28.4.2 Growing phase

The second stage of El Niño is the westward expansion of anomalous conditions appearing off the coasts of Peru and Ecuador. Because of westward expansion of between 50 and 100 cm/s, an exceptionally high rainfall and sea-surface temperature over Christmas Island (20°N, 157°W) can be predicted several months in advance on the basis of Peruvian coastal data. The El Niño phenomenon is in its mature phase between November and January, which is associated with unusually warm surface waters and exceptionally weak trade winds over the tropical Pacific, with the ITCZ displaced further south.

### 28.4.3 Decaying phase

The decaying phase of El Niño is characterized by the return of normal conditions prior to the onset of El Niño conditions off the coast of South America. The anomalous conditions start to decrease a few months after the onset of El Niño. The return to normal conditions to the west of Galapagos Island (0°N, 90°W) does not begin until almost a year later. The decaying phase of El Niño is almost similar to the growing phase in normal conditions.

## 28.5 Theoretical explanation

The El Niño phenomenon can be understood by studying the response of the ocean to the observed meteorological changes and that of the atmosphere to sea-surface temperature variations.

### 28.5.1 Response of the ocean

Along the coastal belt of South America in the Pacific, sea-surface temperature is observed to be minimum, which increases towards both the west and north. These temperature gradients are maintained by the trade winds. If these winds cease, the horizontal redistribution of heat weakens the horizontal gradients. During El Niño/Southern Oscillation (ENSO) the rapid horizontal redistribution of heat apparently causes exceptionally warm surface waters in the upper Pacific. In the initial stages of a typical ENSO the weakening of the meridional component of the trade winds in the eastern Pacific contributes to the appearance of unusually warm surface waters. Normally, the south-east trades south of ITCZ cause coastal upwelling and low sea surface temperatures along the South American coast and upwelling north of the equator, where sea surface temperatures are high (Philander, 1981).

### 28.5.2 Atmospheric response

The warm surface ocean can result in a large release of water vapour to the atmosphere. The horizontal temperature gradients in the tropical atmosphere are small, so that heating of the lower atmosphere causes a rising movement. This makes the air in the surface layers converge on the heated region. In other words, warm surface waters over the eastern tropical Pacific can cause a weakening of the trade winds and even east winds to appear over the western Pacific.

### 28.5.3 Air–sea interaction

As explained above, the warm surface waters over the eastern tropical Pacific weaken the trade winds with a rise in the sea surface temperature in the region of weak winds. The warm water will be transposed eastwards towards the warm anomaly, which causes further expansion of this anomaly in a westward direction.

A part of the warm surface water affects the atmosphere locally in such a way that altered surface wind induces warm oceanic currents, leading to the growth of El Niño. This condition is accompanied by rising air over the water in the south-eastern equatorial Pacific, where the ITCZ is in its southernmost position in February and March. This is why El Niño is born in the spring and its subsequent growth is modulated by the seasonal cycle.

During the mature phase, the convergence of surface winds on the ascending part of the Walker Circulation in the central Pacific is associated with an intensification of the trade winds over the eastern Pacific, where SST starts to fall with the return of normal conditions.

## 28.6 Typical El Niños

### 28.6.1 The El Niño of 1982–1983

This El Niño was one of the worst on record. About 14 countries, from Peru to Australia, were affected by catastrophic floods and drought (Figure 28.2). The usual amount of rainfall in May in Ecuador, for example, is 1.9 inches, but in May 1982, 27 inches recorded over Guayaquil. According to the National Oceanic and Atmospheric Administration (NOAA), the sea

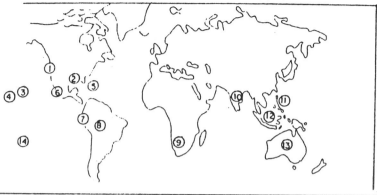

| Phenomena | Location | Period | Victims | Damage |
|---|---|---|---|---|
| **United States** | | | | |
| 1. Storms | Mountain and Pacific States | 9/82–4/83 | 45 + dead | $1 billion |
| 2. Flooding | Gulf states | 12/82–4/83 | 50+ dead | $1 billion |
| 3. Hurricane | Hawaiian Islands | 11/82 | 1 dead | $200 million |
| 4. Drought | Hawaiian Islands | 12/82–4/83 | – | – |
| **Foreign** | | | | |
| 5. Flooding | Cuba | 1/83–3/83 | 8 dead | $170 million |
| 6. Drought | Mexico | 6/82–4/83 | – | $600 million |
| 7. Flooding | Ecuador, northern Peru | 12/82–4/83 | 300 dead | $640 million |
| 8. Flooding | Bolivia | 3/83 | 50+dead | $70 million |
| 9. Drought | Southern Africa | 12/82–4/83 | disease and starvation | $1 million |
| 10. Drought | Southern India and Sri Lanka | 9/82–4/83 | – | $220 billion |
| 11. Drought | Phillipines | 11/82–11/82 | – | $100 billion |
| 12. Drought | Indonesia | 2/82–11/82 | 340 dead | $500 million |
| 13. Drought/files | Australia | 4/82–3/83 | 71 dead/8000 homeless | $3 billion |
| 14. Hurricane | Tahiti | 4/83 | 1 dead | $50 billion |

**Figure 28.2**   *The effect of El Niño, summer 1982 to April 1983 (source: National Oceanic and Atmospheric Administration)*

surface temperature in South America was unusually high (about 10° above normal), which is usually associated with a severe El Niño phenomenon.

Rasmusson said that in Polynesia hurricanes are not expected more than once in 50 years, but in 1982 there were six. The worst disasters occurred in South America and Australia, causing about $8 billion damage (Hilts, 1983). About 1000 animals died and 120 000 people had to flee their homes.

### 28.6.2 *The El Niño of 1986–1987*

During the ENSO event of 1986–1987 (IMD, 1988) the surface oscillation index (SOI), based on pressure data, was negative by March 1986. It decreased from October 1986 to April 1987, i.e. following the failure of the

1986 monsoon. The SOI remained below $-1.0$ during the 1987 monsoon season. This adversely affected the monsoons of 1986 and 1987.

Sea surface temperatures over the equatorial Pacific were higher than normal by about 1°C. With these conditions (negative SST and warm SST), equatorial convection shifted east of its normal position in the Pacific. It is seen that the El Niño events of 1986–1987 began in early 1986, decreased in May to August 1986, and then grew in September–October 1986. It continued thereafter from December 1986 to March 1987 and was prolonged to September 1987 with a slight amplification. The continuous persistence of El Niño over 1986 and 1987 affected the monsoons of both years in which 20% and 48% of India, respectively, suffered drought conditions.

## 28.7  El Niño and monsoon rainfall over India

The impact of a warm episode of El Niño on monsoon rainfall over India has been studied by many workers. Flohn and Fleer (1973) found that several El Niño phenomena were followed by drought conditions during the subsequent monsoon over India. Sikka (1980) identified 22 El Niño years during 1875–1975 and found that in 12 years there was monsoon failure, but in seven years there was none. He found that there were three years of monsoon failure when El Niño did not occur. Rasmusson and Carpenter (1983) identified 25 moderate/strong El Niño years during 1875–1978 and found that the monsoon rainfall over India was generally below normal during these years. Khandelkar and Narella (1984) showed that major drought years were associated with warmer episodes of El Niño events. Parthasarthy and Mooley (1985) examined strong, moderate and weak El Niño years with respect to monsoon rainfall over India and found that mean rainfall increased with a decrease in intensity in El Niño events. Mooley and Paolino (1989) updated the El Niño series to 1984 and identified 27 years of warming phases of the eastern south equatorial (ESE) Pacific. To study the response of Indian monsoon rainfall, the seasonal (June to September) area weighted rainfall over India was calculated. The standardized anomaly (deviation from the mean divided by the standard deviation) was calculated and a drought condition was defined as a standardized anomaly $< -1.28$ and a moderate drought $< -0.50$ but $> -1.28$. Table 28.2 gives the normalized anomaly of Indian monsoon rainfall and years of warming events over the ESE Pacific. It can be seen from the table that a warming event over ESE Pacific is generally followed by drought conditions in India. However, there were three warming years (1948, 1963 and 1976) when India did not experience any drought. In addition, 1987 was a severe drought year and was preceded by an El Niño event (IMD, 1988). IMD (1988) have identified 26 drought years during 1875–1987 on the basis of percentage of area affected by a moderate/strong drought (Table 28.3). Mooley and Paolino (1989) have identified 28 El Niño

**Table 28.2**  Normalized anomaly of Indian monsoon rainfall during years of warming SST over ESE Pacific

| S no. | Year of warming event | Normalized rainfall anomaly |
|---|---|---|
| 1 | 1876 | −0.92 |
| 2 | 1877 | −2.99 |
| 3 | 1883 | −0.04 |
| 4 | 1888 | −0.51 |
| 5 | 1896 | −0.34 |
| 6 | 1899 | −2.70 |
| 7 | 1901 | −0.73 |
| 8 | 1904 | −1.24 |
| 9 | 1905 | −1.65 |
| 10 | 1911 | −1.43 |
| 11 | 1913 | −0.84 |
| 12 | 1918 | −2.46 |
| 13 | 1925 | −0.59 |
| 14 | 1930 | −0.63 |
| 15 | 1939 | −0.76 |
| 16 | 1940 | −0.02 |
| 17 | 1941 | −1.48 |
| 18 | 1948 | +0.24 |
| 19 | 1951 | −1.39 |
| 20 | 1957 | −0.82 |
| 21 | 1963 | +0.04 |
| 22 | 1965 | −1.75 |
| 23 | 1968 | −1.18 |
| 24 | 1969 | −0.28 |
| 25 | 1972 | −2.40 |
| 26 | 1976 | +0.04 |
| 27 | 1982 | −1.42 |
| 28 | 1987 | −1.46 |

*Note*: Long-period (1871–1984) normal rainfall 85.3 cm and SD 8.3 cm

Source: Mooley and Paolino (1989).

years during the same period (Table 28.1). Out of 28 El Niño years, 19 were warming years and nine were non-warming. Analysis of Tables 28.1 and 28.3 reveals that 14 out of 19 warming El Niño years were followed by droughts in the subsequent monsoon years, and there was no drought in the remaining five warming years. There were 19 non-warming El Niño years, out of which seven non-warming years were not associated with droughts and two non-warming years were linked with droughts or delayed droughts. There were three years of drought which were not associated with El Niño at all.

**Table 28.3**   Drought years with percentages of India affected

| S no. | Drought years | Area affected |
|-------|---------------|---------------|
| 1 | 1872 | 59.5 |
| 2 | 1891 | 22.7 |
| 3 | 1899 | 68.4 |
| 4 | 1901 | 30.0 |
| 5 | 1904 | 34.4 |
| 6 | 1905 | 38.1 |
| 7 | 1907 | 29.1 |
| 8 | 1911 | 28.4 |
| 9 | 1913 | 24.5 |
| 10 | 1915 | 22.2 |
| 11 | 1918 | 70.0 |
| 12 | 1920 | 38.0 |
| 13 | 1925 | 21.1 |
| 14 | 1939 | 27.9 |
| 15 | 1941 | 35.5 |
| 16 | 1951 | 35.1 |
| 17 | 1965 | 38.3 |
| 18 | 1966 | 35.4 |
| 19 | 1968 | 21.9 |
| 20 | 1972 | 40.4 |
| 21 | 1974 | 34.0 |
| 22 | 1979 | 34.8 |
| 23 | 1982 | 29.1 |
| 24 | 1985 | 32.7 |
| 25 | 1986 | 19.7 |
| 26 | 1987 | 47.7 |

Source: IMD (1988) report.

## 28.8 Conclusions

On the basis of analysis of long-term data (1875–1987), it can be concluded that El Niño/warming events of SST over the ESE Pacific are generally followed by droughts (below-normal monsoon rainfall) over India. However, the question remains as to why there was no El Niño event in 1979, which was a drought year, and why 1976, a non-drought year, was accompanied by an El Niño event. However, an El Niño event can be used to predict below-normal monsoon rainfall over India in the following year, at least qualitatively.

# References

Flohn, H. and Fleer, H. (1975) *Atmosphere*, **13**, 96–109.

Hilts, J. P. (1983) *Washington Post*.

IMD (1988) *A Report on Scientific Aspects of the Failure of Summer Monsoon Rains over India in 1987*.

Hoskins, B. J. and Simmons, A. J. (1979) *GARP Publ. Ser. No. 22*, **1**, 519–524.

Kilonsky, B. J. and Ramage, C. S. (1976) *J. Appl. Meteor.*, **15**, 972–975.

Khandelkar, M. L. and Neralla, V. R. (1984) *Geophys. Rev. Lett.*, **11**, 1137–1140.

Lamb, H. H. (1977), 'Climate present, past and future', in *Climate History and the Picture*, Methuen, London, p. 835.

Mooley, D. A. and Paolino, D. A. (1989) *Monsoon*, **40**, 4, 369–380.

Parthasarthy, B. and Mooley, D. A. (1980) *WMO Geneva WMO/TD*, No. 87, 265–266.

Phillander, S. G. H. and Pacanowski, K. C. (1981) *Tellus*, **33**, 201–210.

Phillander, S. G. H. (1983) *Nature*, **302**, 295–301.

Quinn, W. H., Zopt, D. O., Short, K. S. and Knoyang, R. T. W. (1978) *Fish. Bull.*, **76**, 663–678.

Ramage, C. S., Adams, C. W., Hon, A. M., Kilonsky, B. J. and Sadler, J. C. (1980) *VHMEF 80–03*, 101.

Ramage, C. S. and Hon, A. M. (1981) *Monsoon Weather Rev.*, **109**, 9, 1827–1835.

Rasmusson, E. M. and Carpenter, T. H. (1982) *Monsoon Weather Rev.*, **110**, 354–384.

Rasmusson, E. M. and Carpenter, T. H. (1983) *Monsoon Weather Rev.*, **111**, 517–528.

Sikka, D. R. (1980) *Proc. Indian Acad. Science (Earth and Planetary Sciences)*, **89**, 179–195.

Wallen, C. G. (1984) 'Present century climate fluctuations in the Northern Hemisphere and examples of their impact', *WMO WCP 87*.

Wyrtki, K. (1975) *J. Phys. Oceanogr.*, **5**, 572–584.

*Part 5*

# Keynote Lectures and Recommendations

# 29 WATER MANAGEMENT IN EGYPT

Essam Radi
*Minister of Public Works and Water Resources,*
*Cairo, Egypt*

It pleases me to express, to all of you, my deepest appreciation for your most welcome attendance at the opening ceremony of this exceptionally important international event. Our seminar tackles one of the most vital issues today for both developed and developing countries. This is because the rapid increase in population and the stress of providing an adequate food supply is the real challenge to all mankind, a fact which requires coordination of effort, intensification of studies and mutual cooperation between all countries and international agencies. This cooperation is undoubtedly the best means (if not the only one) for sound technical planning to enable nations worldwide to overcome the problem of inadequate food for the welfare of their people.

Water has become so precious at present that its careful conservation and efficient development are of foremost concern to all countries. For this purpose, the discussions and recommendations of the seminar today, climatic fluctuations and their impact on water management, will be of direct international interest. Scientists, engineers and decision makers all over the world will be very interested to know the results and recommendations drawn from this important seminar.

I would like to take this opportunity to share with you some issues that will have a positive impact on present-day decision making as far as water management problems are concerned.

First, the occurrence of changes in the climatic conditions, human interference which has disturbed the natural balance of the ecosystem (e.g. forest removal for agricultural expansion, the use of wood for energy production and other kinds of interference which have led to the increase of carbon dioxide in the atmosphere), and the unwise planning of horizontal agricultural and pastoral expansion have all contributed to drought crises, soil deterioration, desertification, and salinization, especially in Asia and Africa. In an attempt to protect the ecosystem from such phenomena, other energy sources, such as solar energy, biomass energy and materials that replace the conventional use of wood and coal for energy production should be seriously considered. Moreover, existing agricultural productivity should be increased by developing new water resources projects and better management practices, including improved irrigation and drainage methods.

Second, drought alone is not the sole reason for the occurrence of famine. The misuse and mismanagement of water resources and short-sighted and improper planning could have even greater negative impacts. Therefore the execution of intensive programmes for rational water use and management has become a crucial issue if we are to guarantee nations' food requirements and diminish the gap between the increase in population and food availability.

Third, in Egypt today, we are aware of the sensitiveness of the ecosystem around the Mediterranean coasts, of coastal pollution, of sea-water intrusion into the groundwater aquifers, and of the possible occurrences of high water tables that affect inhabited areas. All these factors necessitate coordination and teamwork in order to maintain the proper balance of the ecosystem.

Fourth, with both financial support and technical expertise, various international organizations, such as the United Nations System, can assist developing countries to study climatic fluctuations and impacts, eco-systems and water resources management. Such assistance would cer-tainly help to overcome problems and to develop successful plans for joint projects.

Fifth, the development of the means of obtaining climatic data and the regular exchange of these data among co-basin countries of international river basins will facilitate anticipating changes and limiting their potential harmful effects. Environmental challenges and changes in climatic condi-tions require us to give the utmost priority to water resources planning and development projects. This requires the following:

1. Support to facilitate international cooperation and exchange of infor-mation among countries all over the world;
2. Modernization, improvement and proper management of existing irrigation systems, using the most appropriate technology in the fields of irrigation and drainage;
3. Conservation of water and protection from pollution;
4. Development of usable mathematical models to maximize water distri-bution and water control;
5. Study and analysis of natural factors that affect floods; and
6. Optimum use of water to raise the standard of living of individuals.

Being a country of limited water resources, Egypt is aware that it must make a greater effort to improve performance in all the above areas. Egypt has therefore set short- and long-term plans for developing its water resources and for using water in an optimum manner. This was done in cooperation with the United Nations Development Programme and the International Bank of Reconstruction and Development (World Bank), which initiated the Water Master Plan Project for rational water resources development in 1977. The Water Master Plan has the following major objectives:

1. To conduct a comprehensive study of the present and future water resources, including the use of water for irrigation and other purposes;
2. To propose plans and programmes for the efficient use of water for agricultural development, adequate food supply and other necessary uses;
3. To study the change in water quality and the effect of discharging municipal and industrial wastewater into the Nile and the associated irrigation system, and to analyse the proposals for treatment of such wastewater according to international standards to protect both the environment and public health;
4. To determine the investment needed to attain optimum economic development of water resources and to schedule programmes to achieve these goals, and to ensure that such investment funds are available;
5. To improve the operational efficiency of the High Aswan Dam and other hydraulic structures built across the Nile, with the aim of maximizing power generation and control of water distribution, and to develop functional mathematical models necessary for this purpose;
6. To develop flexible programmes and mathematical models that allow incorporation of possible future changes, and hence achieving the policy of rational water development and management;
7. To train people and to raise the efficiency of workers in the field of water resources planning, development and use;
8. To set organizational principles for taking timely decisions concerning water on a national scale through the Water Planning Committee;
9. To draw the major outlines for preparation of water projects and for conducting economic and environmental feasibility studies and evaluations;
10. To update the database and information relevant to water and to modernize the general outline of the water resources developmental plan up to the year 2000; and
11. To improve the operation and management of the water distribution systems by introducing a telemetry system and by applying mathematical models developed for water distribution, together with training of the users on employing this set of models.

I am pleased to report that the Ministry of Public Works and Water Resources has succeeded in drawing up the general outline of the Water Master Plan in accordance with the above-mentioned objectives. It has also accomplished a plan that secured Egypt's water requirements during the drought period in the recent past.

Finally, I believe that, for a country such as Egypt and other arid regions, tackling water problems should be given the highest priority, that water would become increasingly scarce in the field of agriculture, and that the economics of agricultural production will be based to a great extent upon the maximum production obtained per cubic metre of water.

Finally, I would like to thank you all for your interest in and presence at this important seminar. I would also like to express my deepest appreciation to the special contributions of the International Water Resources Association and United Environment Programme which made this seminar possible.

# 30 CLIMATE CHANGE AND WATER MANAGEMENT

Mostafa K. Tolba

*Executive Director, United Nations Environment Programme*
*Nairobi, Kenya*

Science warns us that 70 centuries of human civilization could be threatened in no more than 40 years by climate change and global warming. Since the Villach Conference in Austria in 1985, atmospheric models suggest that our planet faces unprecedented changes in its climatic regimes.

At the global levels there is a large measure of scientific consensus that we are facing global warming. Climate models show this. I am fully aware that models are far from infallible. No matter how sophisticated, they leave gaps in our knowledge. We may need another decade (probably longer) of intense research before we fully understand and can reliably forecast climate changes at regional and national levels. But to await empirical certainty is a brutal logic.

We cannot gamble with our future. There is enough known that we must act, and act now. The lead time for a response is quite long.

We know that the build-up of atmospheric carbon dioxide, methane, CFCs and other greenhouse gases increases the earth's surface temperature by absorbing the outward flow of infrared radiation. We know that, since the Industrial Revolution, carbon emissions have increased by one quarter. Current emissions from fossil fuels now exceed 5 billion tonnes per year and are increasing annually at about 3%.

Atmospheric $CO_2$ is likely to double pre-industrial levels before the end of the next century. Other greenhouse gases – including methane and the CFCs – are increasing at even higher rates. Despite remarkable progress at the international level to eliminate one major group of greenhouse gases – the CFCs, controlled under the Montreal Protocol – evidence shows our planet will continue to overheat. Measurements show that the planet's average temperature has risen by at least 0.5°C in this century. It is estimated to increase by much more – by 1–3°C – in the next 40 to 60 years. That is without precedent.

Climate changes that ended the last great Ice Age were about 5°C: rainfall patterns changed; sea levels rose; forests moved northwards; some species became extinct, others adapted. Such change transformed the ecological face of our planet and shaped the world we know today. But they were gradual. They took place over tens of thousands of years, and allowed life formations to adapt to the changes.

The changes we face threaten an upheaval not over several millennia but in two human generations. It seems unlikely that the eco-systems which nurture all life on earth will be able to adapt quickly enough. The rate of extinction of our planet's biological diversity – already under seige by generations of destructive human activity – will become worse.

The most profound impacts will probably involve water. Sea levels could rise by more than 30 cm. Low-lying regions – indeed, entire islands – could be inundated. More than one-third of the planet's population live near coastal regions. Millions could be affected by increased flooding, more violent storm surges and more frequent hurricanes.

A UNEP-commissioned report released seven months ago suggests that Egypt is one of the countries most at risk. Other countries include The Netherlands, Bangladesh, and the Maldives.

In Egypt, the coastal areas of the Nile Delta are most vulnerable, and they account for 15% of Egypt's total national product and 60% of its fisheries, and house nearly one-quarter of the nation's population. The lower Delta contains large tracts of land below one metre in elevation. Some areas – including coastal lagoons – are below sea level. A small sea-level rise could have profound impacts. It would overwhelm the brackish lakes of the northern Delta, which comprise one-third of the nation's fish catch. Rising sea levels would salinate lakes and aquifers in the lower Nile Delta, thus affecting as much as 20% of Egypt's 35 000 square kilometres of arable land. The lakes of Maryut, Idku, Burrullos and Manzalah would also be engulfed by the sea. Rising sea levels could complicate sewage drainage systems, causing back-up of sewage flows, thereby increasing the risk of infectious diseases in overcrowded cities.

Already, clean water and sewage systems in Egypt are overstressed. The killers in many urban areas are not malaria or sleeping sickness, but ailments like gastroenteritis brought about by poor sanitation. It is estimated that 131 in 1000 born in Egypt die before the age of five, mainly because of diarrhoea. As the largest city in Africa, Cairo's mounting population is straining the city's ageing infrastructures and increasing the threat of typhoid and cholera.

In response, Egypt is undertaking a large engineering project. Reports indicate that more than $1 billion has already been contracted for the 'Greater Cairo Waste Water Project'. An additional $500 million is expected to be spent before project completion in 1991 and the project is expanding to cover the whole country. This is very reassuring. Yet, a word of caution. Planners and engineers should anticipate possible water level changes because of global warming, and ensure that flushing systems can cope with changes to the Nile Delta area.

That is part of a broader challenge to water managers: to anticipate impacts of climate change on the full range of water management and strategies. Climate change challenges a whole range of assumptions behind traditional hydrological analysis: namely, that the past holds the key to the future. Hydrologists assume climate stability. This assumption is of particular importance, given the long life of existing water systems and the long lead times in devising new water projects. The challenge before water managers is this: will you be able to respond quickly enough to climate change? Impacts could be far-reaching.

For example, slight climate variations could affect Egypt's crop of sugarcane and millet in the south and sugarbeet and flax in the north. Climate sets the rhythm of seasonal crop rotations, seedling emergence and late frost hazards. We must gain a clearer picture of the subtle links between crops and climate, and look to anticipated irrigation needs because of changes in rainfall patterns.

Increased rainfall could bring significant wetting in arid and semi-arid regions of Africa, including the Nile Basin. There may be some benefits for agricultural practices in northern Africa.

It is, however, irresponsible at this early date to view climate change in terms of winners and losers. We are not here to gamble, but to manage our natural resources wisely and to pass on to our children a healthy and productive environment.

To do that, we need comprehensive, credible and cooperative water management policies that anticipate possible global warming and consequent climatic change. To date, we know little about the likely runoff patterns in the Nile Basin because of possible rainfall increases. Preliminary models suggest that more rainfall *could* transform ephemeral wadi regimes into intermittent runoff regimes. Such changes *could* create more inland productive lands. That *could* create new homelands for millions of environmental refugees fleeing inundated coastal areas.

Such planning is badly needed, because water scarcity is taking a terrible toll on people everywhere, especially in developing countries. Globally, at least 1.7 billion people do not have adequate drinking-water supplies. At least 3 billion lack access to proper sanitation. In the next 24 hours, while we meet, 25 000 people worldwide will die from waterborne diseases.

Fresh water – not oil – is becoming *the* dominant resource issue in the Middle East and indeed worldwide, which is why we must forge credible strategies for environmentally sound and sustainable water management that meet today's problems and tomorrow's threats due to climate change. We must focus not on what might take place but what *can* be done right now.

Action is beginning. Last month in the Netherlands, over 60 ministers translated political will into coordinated action. Policy options to address climate change are taking shape in three areas:

1. Increase worldwide energy efficiency to reduce carbon emissions by 20% as a start.
2. Implement an aggressive global reforestation policy, whereby a net forest growth of 12 million ha per year is achieved by the year 2000.

3. Establish a workable international funding mechanism to enable developing countries to become full and equal partners in all efforts to address climate change.

Such international cooperation must also entail environmentally sound water management strategies. For example, proper management of the Nile demands coordinated action among the nine countries that share its basin, both upstream and downstream, from the equatorial plateau in Burundi, Rwanda, Zaire, Tanzania, Kenya, Uganda, the southern territories of the Sudan and the Ethiopian highlands.

Cooperation can and is working. Almost four years ago, UNEP launched the Programme of Environmentally Sound Management of Inland Water Resources. Its mandate is to develop inland water strategies, to train experts, provide guidelines for environmentally sound water management and increase public awareness. This programme included UNEP's first action plan developed for the Zambezi River Basin, Africa's fourth largest river, shared by more than 20 million people in eight countries. Countries are working together to control serious pollution problems from sewage, industrial pollution and mining. More recently, UNEP began preparations for the Lake Chad basin programme, to address serious impacts of climate variability by way of water resource management. In less than 15 years, because of climatic variability, water levels in Lake Chad have dropped considerably.

In a new programme being prepared for the Aral Sea in the Soviet Union, action is taking shape to combat increased industrial pollution and over-exploitation of this large water body.

UNEP is working to include a full range of climate change predictions into these and future Inland Water Management programmes. There are at least *seven* interrelated factors that should set the dimensions of water management strategies in anticipated climate change scenarios:

1. Changing evapotranspiration, including increased air humidity, and increased cloudiness which could destabilize water balances in basins.
2. Increased uncertainty and risk in the design of water resource development projects. Current projects assume climatic and precipitation predictability. Predictability is changing to uncertainty. Climate change could alter hydrological regimes. Floods could become more severe. Project designs must account for such great variations.
3. Water resource managers must start thinking in longer time frames. Water designs should assess the risks of megaprojects like dams, the benefits of smaller, local water projects, in uncertain climatic conditions. Smaller projects could act as an interim stage, at least until we know more about regional impacts of climate change.
4. Water management strategies should anticipate higher instability in erosional and sedimentation processes.
5. Climate change could alter biochemical oxygen demand in coastal areas and endanger coastal areas because of salt water intrusion.

6. Altered sea levels could flood city canalization and disrupt sewage treatment systems.
7. Climate change could affect the lithosphere, with profound impacts on groundwater aquifers, soil moisture, fresh water supplies and irrigation schemes.

I have outlined above some of the challenges facing the designers of water management strategies. Sweeping challenges, because proven assumptions of water management strategies may be entirely ineffective in new climatic regimes. During the last four days of this meeting, you have worked together to help redefine and focus environmentally sound water management strategies within new climatic conditions. We badly need your work and that of groups like you all over the globe. I thank you for what you are doing and urge you to continue your research and policy formations to meet the challenges of climate change.

# 31 SEMINAR REPORT AND RECOMMENDATIONS

The International Seminar on Climatic Fluctuations and Water Management was held at the Meridien Hotel in Cairo, Egypt, from 11 to 14 December 1989. It was sponsored by the Water Research Centre, Ministry of Public Works and Water Resources, Government of Egypt, the International Water Resources Association, and the United Nations Environment Programme, with the support of the USAID/WRC Project.

The Seminar was attended by some 150 participants from 38 countries and many international organizations and 70 papers were received for presentation and discussion. His Excellency Essam Rady, Minister of Public Works and Water Resources, Government of Egypt, and Dr Mostafa Kamal Tolba, Executive Director, United Nations Environment Programme, Nairobi, Kenya, addressed the Seminar.

Currently, there is considerable consensus among the international scientific community that man is facing global warming due to increasing levels of atmospheric carbon dioxide, methane, CFCs and other greenhouse gases. The perceived wisdom is that the average global temperature has increased by between 0.5°C and 0.7°C since the turn of the century (or, according to some, since the 1860s). Current estimates from models indicate that they may increase by 1–3°C during the next 40–60 years. However, since there are considerable gaps in our knowledge and data availability, the reliability of such models leaves much to be desired. Accordingly, these predictions can, at best, be considered as possible scenarios. At our present state of knowledge, even if it is assumed that the global warming will take place on the scale and over the time period predicted (which is far from certain), it cannot be translated into specific changes in temperature and precipitation with any degree of reliability over a watershed for the planning, management and operation of specific individual water projects.

The Seminar papers were presented and discussed in four technical sessions: Monitoring, Forecasting and Analytical Procedures; Impacts of Climatic Fluctuations; Water Resources Planning and Management and Case Studies and Reports. A field trip was organized to the Salhiya and Ismailia areas. The participants were taken to the Salhiya Agricultural Project, where they received an overall briefing on the project and then visited the main pumping station, sprinkler and drip-irrigation systems and greenhouses.

After discussion of all the papers presented on climatic fluctuations observed in the past and the potential impacts of climatic changes that may occur in the future and their implications for water planning and management, the participants in the International Seminar made the following recommendations:

1. There is an urgent need to establish and/or expand reliable and cost-effective data-collection systems for hydrometeorological data, especially for rainfall, runoff, evapotranspiration and temperature, in most developing countries. A reliable long-term database is essential not only for efficient planning, management and operation of water resource systems but also to determine in the future the extent of climatic changes (if any) and how to deal with them effectively. Data collected should be easily available to all parties requiring that information. This is especially important for international river basins, where exchange of data between co-basin countries is essential for optimal water management.

2. There is an urgent need to critically re-examine the existing methodologies used for water planning and management. On the basis of climatic fluctuation observed in many countries and watersheds during the past 30–70 years it is becoming increasingly clear that many of the present hydrological techniques used to generate synthetic streamflow for design of water resources projects leave much to be desired. Even when 30 years of continuous hydrological time-series data are available, it is currently assumed that they define hydroclimatic averages. Current analyses from many parts of the world indicate that this assumption may not necessarily be accurate. For many water projects in developing countries, often even 30 years of data are not available. Under these conditions, elasticity of water projects to deal with unexpected risks from climatic fluctuations could become an important consideration. Accordingly, the development of new and more effective methodologies which would enable us to plan, design and operate water projects more efficiently than at present has become an urgent necessity.

3. Increasing attention should be paid to improve existing monitoring and forecasting methods which could facilitate more effective consideration of climatic fluctuations in water management. Equally, facilities available for such activities in developing countries should be significantly increased. International organizations should make a special effort to assist developing countries to establish effective forecasting centres and also to improve the operational capabilities of existing ones.

4. Special attention and support should be given to international river basins and aquifers in terms of their more efficient management due to climatic fluctuations. Within this context, authorities in charge of international water bodies should be provided with increasing support to improve their capacity to:
   (a) Monitor and predict patterns of rainfall, runoff, temperature, and evapotranspiration;

(b) Plan, manage and operate water projects efficiently in the light of potential climatic fluctuations;

(c) Analyse and forecast environmental impacts of climatic fluctuations and steps that can be taken to effectively manage such impacts; and

(d) Estimate environmental, social and economic costs due to both natural changes and planned hydraulic interventions.

5. The International Water Resources Association should consider the establishment of a network of researchers studying climatic fluctuations and water resources issues as a Specialty Group, with the full support and active collaboration of UNEP and other appropriate international organizations such as WMO and FAO. Such a Group could:

(a) Compile, maintain and disseminate a comprehensive, annotated bibliography on climatic fluctuations and water management;

(b) Critically evaluate the present techniques of assessing time series, and develop new and more improved methodologies;

(c) Provide a better understanding to identify and integrate the effects of climatic fluctuations and possible future climatic changes;

(d) Encourage dissemination of climatic change scenarios derived from GCMs, historical analyses and other methods; and

(e) Examine the need for an international institute on climate and water management and, if considered desirable, explore the possibilities of establishing such an institute.

6. Training of water resources professionals on how to effectively incorporate climatic fluctuations into planning, management and operation of water resources systems is needed. The training programmes should emphasize practical and operational aspects of water projects. At present, such effective training programmes do not exist. Universities, training centres and international organizations should work together to develop such training courses.

7. International organizations such as the United Nations Environment Programme, the FAO, the International Irrigation Management Institute, the WMO and other appropriate organizations should give more emphasis on how to effectively incorporate climatic fluctuations into water management. They should also assist selected developing countries to establish new centres for studies on climate and water management.

8. The Water Research Center, the International Water Resources Association and the United Nations Environment Programme are to be complimented on organizing a very effective seminar on climatic fluctuations and water management. The Seminar greatly facilitated knowledge and information transfer between water and climate professionals from numerous countries. Selected important papers from the seminar should be published as soon as possible and distributed widely to potential users. Since it is very difficult for water and climate professionals in developing countries to obtain the latest information on this important topic, UNEP and appropriate international organiza-

tions should be requested to support complimentary distribution of the book resulting from this seminar to potential users in developing countries.

A similar international seminar should be organized in 1994 to consider and review new developments during the next three years. Every attempt should be made to make the 1994 seminar multi-disciplinary.

# INDEX